Fundamentals of Flow Measurement

An Independent Learning Module from the Instrument Society of America

FUNDAMENTALS OF FLOW MEASUREMENT

By Joseph P. DeCarlo
THE FOXBORO COMPANY
Foxboro, Massachusetts

INSTRUMENT SOCIETY OF AMERICA

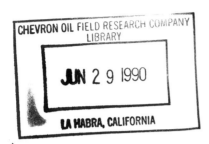

Copyright © Instrument Society of America 1984

All rights reserved

Printed in the United States of America

No part of this publication may be reproduced, stored in a retrieval system, or transmitted, in any form or by any means, electronic, mechanical, photocopying, recording or otherwise, without the prior written permission of the publisher.

INSTRUMENT SOCIETY OF AMERICA
67 Alexander Drive
P.O. Box 12277
Research Triangle Park
North Carolina 27709

ISBN 0-87664-627-5
L.C. # 83-12686

Library of Congress Cataloging in Publication Data
DeCarlo, Joseph P., 1937–
   Fundamentals of flow measurement.

  (An Independent learning module from the Instrument Society of America)
  1. Flow meters. I. Title. II. Series: Independent learning module.
TJ935.D3 1983   681'.2   83-12686
ISBN 0-87664-627-5

Editorial development and book design by Monarch International, Inc.
Under the editorial direction of Paul W. Murrill, Ph.D.

Production by Publishers Creative Services, Inc.

Table of Contents

Preface — vii

Comments — vii

UNIT 1 **Course Overview and Introduction**
- 1-1. Purpose — 3
- 1-2. Audience and Prerequisites — 3
- 1-3. Study Material — 4
- 1-4. Organization and Sequence — 4
- 1-5. Course Objectives — 5
- 1-6. Course Length — 5

UNIT 2 **Classification of Flowmeters**
- 2-1. Approaches to Flow Measurement and Classes of Flowmeters — 9
- 2-2. Extractive Energy Approach — 11
- 2-3. Additive Energy Approach — 18

UNIT 3 **General Flow-Measurement Terminology**
- 3-1. Flowmeter Physical Terms — 27
- 3-2. Fluid-Related Terms — 29
- 3-3. Flowmeter Performance Terminology — 34
- 3-4. Flowmeter Calibration Terminology — 38

UNIT 4 **Theory of Differential Pressure Flowmetering Devices**
- 4-1. Incompressible Flow — 45
- 4-2. Compressible Flow — 50
- 4-3. Laminar Flow — 52
- 4-4. Turbulent Flow — 55

UNIT 5 **Head-Producing Flowmeters I (Conventional)**
- 5-1. Orifice Plates — 62
- 5-2. Venturi Tubes — 73

UNIT 6 **Head-Producing Flowmeters II (Special)**
- 6-1. Pipe Elbow or Centrifugal Flowmeters — 87
- 6-2. Target Flowmeter — 90
- 6-3. Pitot and Pitot-Static Probes — 92
- 6-4. Area Flowmeters — 97
- 6-5. Linear-Resistance Flowmeters — 100

UNIT 7 **Head-Producing Flowmeters III (Open Channel)**
- 7-1. Weirs — 107
- 7-2. Flumes — 117

UNIT 8 **Pulse Producing Flowmeters**
- 8-1. Positive-Displacement-Type Flowmeters — 129
- 8-2. Current-Type Flowmeters — 136
- 8-3. Fluid-Dynamic-Type Flowmeters — 143

UNIT 9 **Powered Flowmeters**
- 9-1. Magnetic Flowmeters — 157
- 9-2. Ultrasonic Flowmeters — 163
- 9-3. Thermal Flowmeters — 170

UNIT 10 **Special Techniques**
    10-1. Tagging (Flow Marker or Tracer) Technique   183
    10-2. Deflection   192
    10-3. By-Pass (Flow Ratio Technique)   195

UNIT 11 **Mass-Flow Measurement**
    11-1. Inferential Mass-Flow Measurement   204
    11-2. True Mass-Flow Measurement   208

UNIT 12 **Flowmeter Selection**
    12-1. Flowmeter Selection Process   223
    12-2. Example of Flowmeter Selection   231

APPENDIX A: **Suggested Readings and Study Materials**   241

APPENDIX B: **Solutions to All Exercises**   245

APPENDIX C: **Glossary of Flow Measurement Terminology**   263

Index   271

Preface

ISA's Independent Learning Modules

This is an Independent Learning Module (ILM) on the subject of flow measurement, it is part of the ISA Series of Modules on *Fundamental Instrumentation*.

The ILMs are the principal components of a major educational system designed primarily for independent self-study. This comprehensive learning system has been custom designed and created for the ISA to more fully educate people in the basic theories and technologies associated with applied instrumentation and control.

The ILM System is divided into several distinct sets of modules on closely related topics; such a set of individually related modules is called a series. The ILM System is composed of:

The ISA Series of Modules on *Control Principles and Techniques*.

The ISA Series of Modules on *Fundamental Instrumentation*.

The ISA Series of Modules on *Unit Process and Unit Operation Control*.

The principal components of a series are the individual ILMs such as this one. They are especially designed for independent self-study; no other texts or references are required. The unique format, style, and teaching techniques employed in the ILMs make them a powerful addition to any library.

Comments about This Volume

This ILM, *Fundamentals of Flow Measurement*, is designed to be an introduction to the basic principles and practices used in the various flow-measurement techniques. Mathematical development is kept to a minimum and only used to emphasize the underlying physics and theory of operation for a particular flow-measurement device. Consequently, the mathematics presented deals only with fundamentals and the actual working equations are avoided in most cases. In addition, due to the complex nature of gas flow measurements, the subject is only briefly introduced and methods and practices are not discussed in great depth.

Most important is the understanding of the classification and characterization of the approaches, classes, types, and actual devices used in flow measurement. Once the classification of the various devices is understood, the flowmeter selection process for a particular application becomes much easier to understand.

Unit 1: Course Overview and Introduction

UNIT 1

Course Overview and Introduction

Welcome to ISA's Independent Learning Module on flow measurement. This unit provides you with an overview of the course and the information you need to take the course.

Learning Objectives — When you have completed this unit, you should:

A. Know the nature of material to be presented.

B. Understand the general organization of the course.

C. Know the course objectives.

1-1. Purpose

The purpose of this ILM is to convey to the student a comprehensive knowledge of flow-measurement methods through documentation that is logically categorized and readily utilized in practical situations. Course structure is purposely divided into segments representing certain classes of flow-measurement methodology and definitive types of flowmeters. Using this approach, the student must determine the flowmeter class, or method, which promises the best solution to the problem. Once the class is defined, additional effort is required to determine the best type of flowmeter within that class and to size the chosen flowmeter to meet the requirements.

1-2. Audience and Prerequisites

This ILM is designed for persons who wish to work on their own at their own pace and desire to learn the basics of flow-measurement theory and application. The material should be useful to senior technicians, first-line supervisors, and engineers who are concerned with flow measurement, or to students in technical schools, colleges, or universities who wish to gain some insight into the principles and practices of flow measurement.

No particular prerequisites are required other than the interest and motivation to complete the course. Some mathematical development is necessary. However, it is used more as an explanation of the physics of the flow-measurement approach rather than for the development of the working equations of a particular flowmeter. Mathematics should not be the barrier that prevents full understanding of flow measurement. This ILM is designed to eliminate the mathematical barrier often found in texts on flow measurement by avoiding the working equations while emphasizing the connection between the physics of operation and the fundamental equation that describes the phenomenon.

1-3. Study Material

This textbook is the only study material required in this course. It is one of ISA's several Independent Learning Modules on fundamental instrumentation designed as an independent, stand-alone textbook. It is uniquely and specifically structured for self-study. A list of suggested readings in Appendix A provides additional references and study materials.

The student also may find it helpful to study other Independent Learning Modules available from ISA. In conjunction with flow, simultaneous study topics might be pressure, level, and temperature measurement.

1-4. Organization and Sequence

This ILM is divided into twelve separate units. The next two units are designed to introduce the student to the general classes and types of flowmeters and the terminology associated with flow measurement and flowmeters. Units 4 through 8 are devoted to a flow-measurement approach called "energy extractive" which forms the foundation of some of the oldest and most widely used means of flow measurement. Unit 9 is devoted to a flow-measurement approach called "energy additive" which forms the foundation of some of the newest employed means of flow measurement. Unit 10 covers special techniques most often applied in difficult flow-measurement situations. Unit 11 covers mass-flow measurement. The last unit is a general review in the form of a methodology for flowmeter selection. It outlines the thought process, calculations, and cost considerations involved in flowmeter selection.

Each unit is designed in a consistent format with a set of specific *learning objectives* stated in the very beginning of the unit. Note these learning objectives carefully; the material which follows the learning objectives will teach to these objectives. The individual units often contain example problems to illustrate specific concepts. At the end of the unit you will find student exercises to test your understanding of the material. All student exercises have solutions contained in Appendix B.

1-5. Course Objectives

When you have completed this entire Independent Learning Module, you should be able to:

A. Describe the basic approaches to flow measurement and list the different classes of flowmeters.

B. Converse comfortably using flow-measurement terminology.

C. Explain the basic theory of operation of the different classes and types of flowmeters.

D. Choose the correct approach, class, type, and particular device that best satisfies your requirements.

E. Size a particular flowmeter to satisfy the requirements of your installation.

In addition to these overall objectives, each unit contains a specific set of learning objectives — highlighted as in this first unit — that are intended to be very specific in order to help direct your study of that particular unit.

1-6. Course Length

The basic idea of the ISA System of Independent Learning Modules is that students should be able to learn best if they proceed at their own pace. As a result, there will be significant variation in the amount of time taken by individual students to complete this ILM.

You are now ready to begin your in-depth study of flow measurement. Please proceed to Unit 2.

Unit 2: Classification of Flowmeters

Unit 2

Classification of Flowmeters

There are many flowmeters on the commercial market today, designed for a variety of industrial situations and employing several measurement methods. This chapter contains a listing of the most important types of flowmeters classified by the physical principles governing their operation.

Learning Objectives — When you have completed this unit, you should:

A. Have a general understanding of the classes and types of flowmeters.

B. Be able to recognize a type and class of flowmeter.

C. Have some knowledge of the operating principle and use of each class of flowmeter.

2-1. Approaches to Flow Measurement and Classes of Flowmeters

In the most simplistic terms, all flow-measurement instruments fall into two categories which are, in reality, "approaches" to flow measurement. These categories are: the *extractive energy approach* and the *additive energy approach*, as depicted by the caricature sketch in Fig. 2-1.

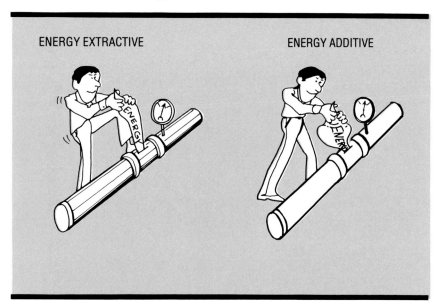

Fig. 2-1. Approaches to Flow Measurement

What is energy? Energy may be defined as the ability of an agent to do work. For example, a man can do work and, therefore, possesses energy. Steam possesses energy because it is able to drive the piston of a steam engine. Water possesses energy because it drives the turbine in a hydroelectric plant. Fluid in motion possesses energy by virtue of its motion. This type of energy is called kinetic energy. Fluid at rest also possesses energy by virtue of its position or height above a place to which the fluid could possibly flow. This type of energy is called potential energy.

Energy may change from potential to kinetic and back to potential. In a flow line, the flowing fluid possesses both potential and kinetic energy. The faster the fluid flow, the more energy is transformed from potential to kinetic. During the transformation of energy states (potential to kinetic), some energy is expended in working against friction. It is said to be wasted or, in reality, rendered unavailable for useful purposes. It is not destroyed, however, for it is converted to heat, another form of energy.

Because fluid flowing in a pipeline possesses energy, both potential and kinetic, one approach to measuring the flow rate of the fluid in the pipeline may be considered *energy extractive*. This approach involves the placement of a device in the flowing stream of fluid that either changes some of the potential energy into kinetic energy and, in doing so, some energy is "extracted" or lost in the form of pressure; or "extracts" a certain amount of kinetic energy in the form of work done by the fluid on an object placed in the flow, such as a rotor or turbine.

A second basic approach to flow measurement is termed *energy additive*. In this approach, some outside source of energy is introduced to the flowing fluid and either the effect on the introduced source by the fluid or the effect on the fluid by the introduced source is monitored. Usually, additive energy approaches to flow measurement are nonintrusive or present very low blockage to the flow, which is an inherent advantage in the basic philosophy of this approach to flow measurement.

Let us first consider in more detail the *extractive energy approach*, since this is the oldest and most commonly used.

2-2. Extractive Energy Approach

The oldest class of flowmeter using the extractive energy approach is the differential-pressure-producing class of flowmeter. There are many types of flowmeters in this class, the most conventional being the orifice, nozzle, and venturi shown in Fig. 2-2. A graphic display of the operating principles of this general class of differential-producing flowmeters is presented in Fig. 2-3.

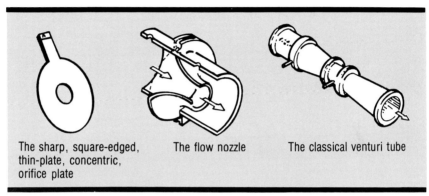

The sharp, square-edged, thin-plate, concentric, orifice plate

The flow nozzle

The classical venturi tube

Fig. 2-2. Conventional Types of Differential-Pressure-Producing Class of Flowmeters

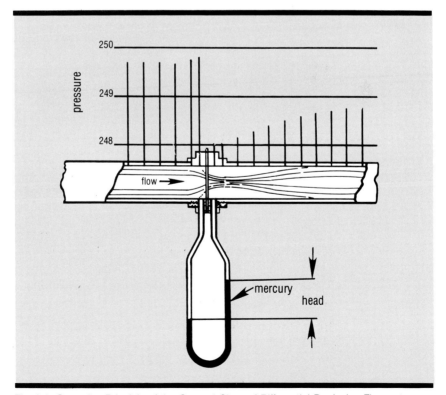

Fig. 2-3. Operating Principle of the General Class of Differential-Producing Flowmeters

Note that the constriction placed in the flow line (orifice in this case) produced a change in pressure on the wall of the pipe and across the orifice. This change in pressure as measured by the manometer shown, is proportional to the rate of fluid flow. The pressure rise (recovery) after the orifice never reaches the pressure level ahead of the orifice; this pressure loss is an indication of the energy extracted from the pipeline resulting from the presence of the orifice plate.

A point of interest, both historical and factual, in today's usage of flowmeter terminology is that the differential pressure generated by the orifice can produce a *head* of liquid, namely mercury in the case of the manometer shown. *Head* is used here to refer to the height of liquid produced in a manometer arrangement, from which the differential pressure may be quantized. From this viewpoint, the general class of differential-pressure-producing flowmeters is commonly referred to as *head-class meters*. More descriptive and correct usage of the term *head-class* flowmeter is applied to *open-channel* flowmeters such as the weir depicted in Fig. 2-4. In this type of flowmeter the flowing liquid changes its height or *head*, and the height measurement is the parameter related to flow rate. In any case, the term *head class* is used in this text to refer to the general class of differential-pressure-producing flowmeters.

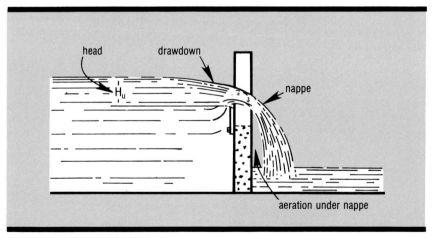

Fig. 2-4. Free Flow over a Weir

The foundation for the theory of the head meter was laid about the start of the seventeenth century with the development of two concepts: the volume rate of flow was equal to the velocity times the area, and the flow rate through an orifice varies as the square root of the *head*, differential pressure. Finally, in 1738,

John Bernoulli developed the theorem on which the hydraulic equations for head meters are based. In 1797 Giovanni Battista Venturi published the results of his basic work on the principles of the venturi tube (Ref. 1).

The nozzle, the venturi tube, and the orifice plate are considered conventional-type devices in the head-producing class of flowmeters. We use the word *conventional* here to distinguish these three most commonly used flowmeters from a group-type called *special*. The special group includes such devices as the target meter, elbow meter, area meters, pitot probes, and linear-resistance meters, shown in Fig. 2-5.

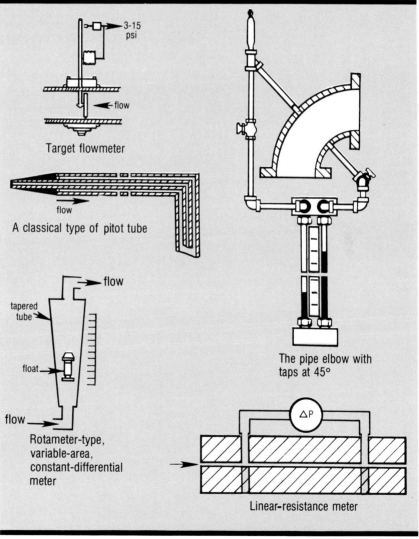

Fig. 2-5. Special-Type Devices in the Head-Producing Class of Flowmeters

Another type of head-class flowmeter is called the *open-channel type*. Like the conventional type, they are steeped in history and tradition. Particular devices in the open-channel type of head-class flowmeters are the weir and flume.

Another class of flowmeters using the *extractive energy approach* to flow measurement is the *pulse-producing class*. This class includes three basic types of flowmeters: *positive-displacement type*, *current type*, and *fluid-dynamic type*.

Positive-displacement-type flowmeters are placed in the *pulse class* because each discrete volume of fluid is represented by a pulse or countable unit. Summing the pulses results in a total quantity of flow. Of the several types developed, the nutating-disk is one of the oldest. In this type meter, Fig. 2-6, a circular disk is attached to a spherical center and restrained from rotating. It is allowed to nutate (a floppy or bobbing motion), however, so that the liquid entering the inlet port moves the disk until the liquid discharges from the outlet port. A mechanical counter indicates the number of cycles the disk nutates, and each cycle represented by a count or pulse is proportional to a specific quantity of flow. Another very popular positive-displacement type of flowmeter is the *oval gear meter*, Fig. 2-7. In this design, two oval-shaped gears with close-fitting teeth are synchronized such that a fixed quantity of liquid is put through for each complete rotation. The shaft rotation is monitored so that the number of revolutions in a unit time represents a specific flow rate, and the total number of revolutions indicate the total integrated flow quantity.

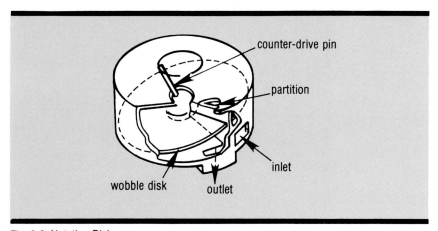

Fig. 2-6. Nutating Disk

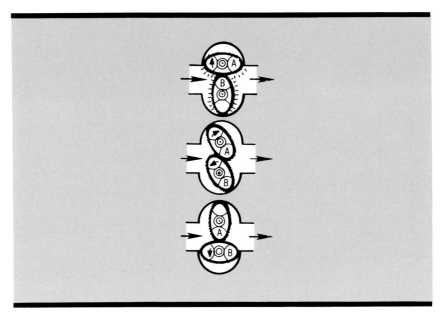

Fig. 2-7. Oval Gear

The earliest form of *current-type* flowmeter was disclosed by Benjamine Gottlob Hoffman in Hamburg around 1790 in a booklet describing a meter for measuring flow of air and water invented by Reinhard Woltman. Generally, the design, Fig. 2-8, consisted of a propeller or windmill-like vanes which were pushed or rotated by the water current.

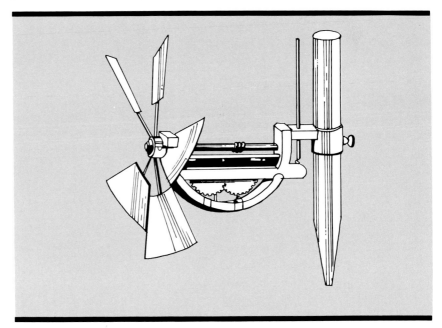

Fig. 2-8. Original Form of Woltman Meter

The propeller meter used today bears a close resemblence to the Woltman current meter, and the modern-day turbine meter, Fig. 2-9, is basically a refined propeller meter, for application in closed conduits.

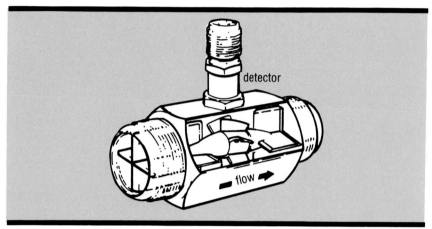

Fig. 2-9. Turbine Meter for Liquids

A very different type of pulse-class flowmeter is the *fluid-dynamic* type in which the underlying principle of operation lies in the generation of an oscillating motion in the fluid. The oscillating motion is then monitored in the form of discrete pulses to arrive at flow rate and total flow quantity. The attraction of this type of pulse-class meter is that it is the only type of pulse-class meter having no moving parts. Basically, there are two devices in the fluid-dynamic type of pulse-class flowmeters. They are the vortex flowmeter and the fluidic oscillator, shown in Fig. 2-10.

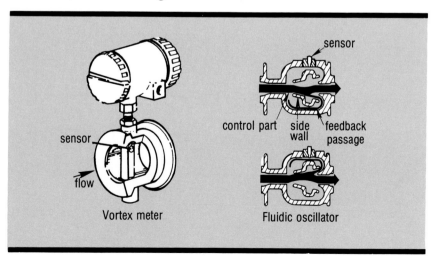

Fig. 2-10. Fluid-Dynamic Type of Pulse-Class Flowmeter

Vortex-generating body shapes and the relationship between the frequency of generated vortices and the fluid velocity have been known since about 1878 when V. Strouhal observed that the frequency of eddies or vortices produced behind a bluff body increased with flow rate in a linear fashion.

The fluidic oscillator device is based on the use of a fluid-dynamic phenomenon known as the "Coanda" effect which, simply stated, is the tendency of a fluid jet in the vicinity of a wall to adhere to that wall. If the geometric shape of the meter body is such that the initiating flow attaches itself to one of two walls (the "Coanda" effect) and a small portion of the flow is feedback to a control port, the feedback flow acting on the main flow will divert the main flow to the opposite wall. Repeating the same feedback action results in a continuous self-induced oscillation in the flow between sidewalls. The frequency of oscillation is linearly proportional to the fluid velocity just as for the vortex-generating flowmeter.

As a brief review of the preceding discussion on the *extractive energy approach* to flow measurement, study the family tree of the *head* and *pulse* classes of flowmeters provided in Fig. 2-11. Note the breakdown of each class into the different types of flowmeters and the specific devices within those types.

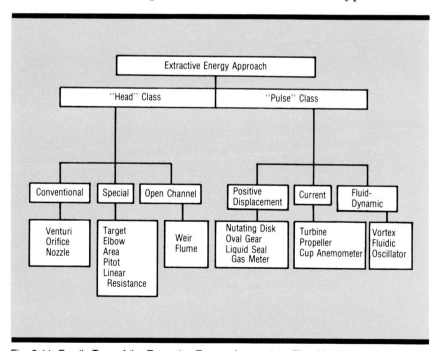

Fig. 2-11. Family Tree of the Extractive Energy Approach to Flow Measurement

2-3. Additive Energy Approach

Remember that the *additive energy approach* to flow measurement is characterized by the use of some outside energy source, and either the effect of the flowing fluid on the introduced energy source or the effect on the fluid by the source is observed. Most often, the outside energy source is generated through the use of electrical power; however, nuclear power is used also.

The three most prominent classes of flowmeters using the *additive energy approach* to flow measurement rely on electrical power and introduce the energy source to the flow in the form of magnetism, sound, and heat. Consequently, the three classes are defined as the *magnetic class*, the *sonic class*, and the *thermal class*. We will not treat *nuclear* as a most often used or prominent class of flow measurement because its use is limited more to a special technique of flow measurement to be treated in Unit 10.

Consider first the *magnetic class* of flowmeter. There are two types of flowmeters in the *magnetic class*, the AC ("alternating current type") and the DC ("direct current type"). Both types use electricity to power an electromagnet which introduces the magnetic energy source into the flowing fluid.

The principle of operation of the magnetic flowmeter was discovered in 1831 by Michael Faraday, a British physicist, who discovered that a voltage was induced in a circuit by providing a relative motion between the circuit and a magnet.

The schematic representation for a magnetic flowmeter is illustrated in Fig. 2-12. An electrically conducting liquid, carried in a nonconducting, nonmagnetic tube or pipe, passes between the poles of an electromagnet. The flow of liquid is perpendicular to the lines of magnetic force, and the ions contained in the liquid, under the action of the magnetic force, give up their charges to the measuring plates or electrodes. This action produces an electromotive force proportional to the liquid flow rate. To circumvent the problem of electrolysis, an electromagnet is used rather than a permanent magnet. The electromagnet is powered either by alternating current or a pulsed direct current. Consequently, the two flowmeter types in the *magnetic class* we will be discussing in more detail later are the AC type and the pulsed DC type.

Unit 2: Classification of Flowmeters 19

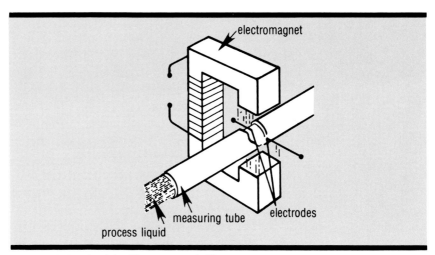

Fig. 2-12. Principle of the Electromagnetic Flowmeter

In the *sonic class* of flowmeters, there are two important types that deserve discussion; the *time-of-flight (TOF)*, or differential time, and the *doppler*. Before going too far with our discussion of the types of *sonic-class* flowmeters, let's go back to the basics of the additive energy approach to flow measurement. For most all of the *sonic-class* flowmeters, electrical energy is used to excite a piezoelectric or crystal-type material to its resonant frequency. This resonant frequency is transmitted in the form of a wave, traveling at the speed of sound, in the material the crystal is touching. If we introduce this wave into a fluid flowing in a pipe in such a way that the sound wave travels against the flow in one direction and with the flow in the opposite direction, shown in Fig. 2-13, the difference in transit time of the wave is proportional to the fluid flow rate. This is so because when traveling against the fluid flow, the sound wave is slowed by a small amount, and when traveling with the fluid flow, the sound wave velocity is increased.

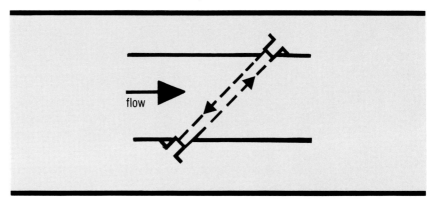

Fig. 2-13. Contrapropagating Transmission (upstream-downstream)

The Doppler effect, discovered in 1842 and commonly used today in radar systems and depth gauges (Sonar), is useful in medical and biological studies, some industrial pipeline applications, and marine applications. A real-life demonstration of the Doppler effect is heard when an on-coming horn-blowing car or train passes a stationary observer. The tonal quality (frequency) heard as the train approaches is quite different than that heard as the train departs from the observer. Similarly, when an ultrasonic beam is projected into an inhomogeneous fluid, some acoustic energy is scattered back toward the transducer. Because the fluid is in motion with respect to the transducer and the scattered sound moves with the fluid, the received signal differs from the transmitted signal by a certain frequency referred to as the Doppler shift-frequency. This shift-frequency is directly proportional to the fluid flow rate. A schematic representation of the ultrasonic *Doppler-type* of *sonic-class* flowmeter is shown in Fig. 2-14.

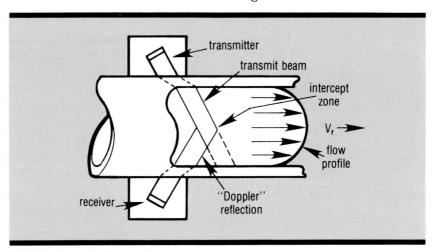

Fig. 2-14. Ultrasonic Doppler Type

The entire class of the sonic additive energy approach to flow measurement will be explored in great detail later in this study course.

In the *thermal class* of flow measurement, the flow rate is measured either by monitoring the cooling action of the flow on a heated body placed in the flow or by the transfer of heat energy between two points along the flow path. Here, the additive energy is heat, usually produced by an electrical source. Two types of flow-measurement devices in the *thermal class* of flowmeters are the *thermo-anemometer* and the *calorimetric flowmeter*, shown in Fig. 2-15.

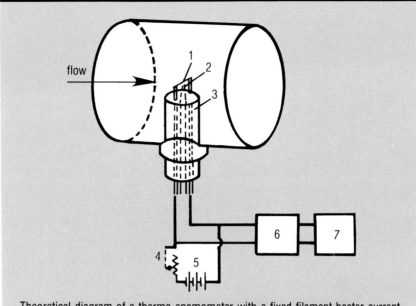

Theoretical diagram of a thermo-anemometer with a fixed filament heater current.
1 — heated filament; 2 — low-inertia thermocouple; 3 — body of thermo-anemometer; 4 — regulating resistor; 5 — power supply; 6 — amplifier; 7 — recording instrument.

Thermo-Anemometer Type

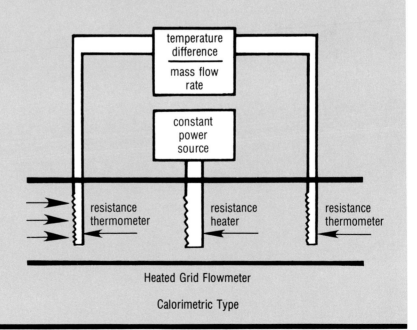

Heated Grid Flowmeter

Calorimetric Type

Fig. 2-15. Thermal Class of Flowmeters

The working principle of the *thermo-anemometer* is simply based on the monitoring of electrical parameters that change due to the removal of heat by the flow from a heated body immersed in the flow. In contrast, the *calorimetric flowmeter* operates on the principle of measuring the rate at which the flow transfers heat energy from a heat sensor.

As a brief review of the *additive energy approach* to flow measurement, study the family tree of the *magnetic, sonic,* and *thermal* classes of flowmeters provided in Fig. 2-16. Note the breakdown of each class into different types of flowmeters and the specific devices within those types. The smaller family tree, in comparison to the *extractive energy approach*, is an indication of the relative newness of the *additive energy approach* to flow measurement. It should not be misinterpreted as the less favorable approach to flow measurement.

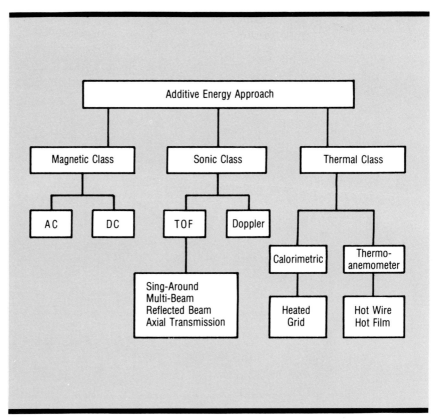

Fig. 2-16. Family Tree of the Additive Energy Approach to Flow Measurement

Exercises

The following types of flow-measurement devices and meters were discussed in this chapter:

Weir	Oval gear meter
Target meter	Flume
Orifice plate	Fluidic oscillator
Pitot probe	Propeller
Magnetic (AC or DC)	Nutating disk
Area meter	Elbow meter
Vortex	Thermo-anemometer
Liquid seal meter	Turbine
Cup anemometer	Nozzle
Time-of-flight	Calorimetric
Linear resistance	Venturi

2-1. *Which of the above meters use the extractive energy approach?*

2-2. *Separate the energy extractive meters into head-class and pulse-class meters. Which meters have a frequency output?*

2-3. *Which meters use the additive energy approach? Classify these meters by the type of energy added to the fluid.*

2-4. *"Conventional" head-class meters are very common in industrial applications. Which of the above are so classified?*

References

[1]*A History of Flow Measurement by Pressure-Difference Devices*, Publication 1010/460, George Kent Limited.

Unit 3: General Flow-Measurement Terminology

Unit 3

General Flow-Measurement Terminology

This unit is intended to instruct you in the terminology (jargon) of the flowmetering community. Even more important than the terminology are the ways flowmeter performance is specified.

Learning Objectives—When you have completed this unit, you should:

A. Have a conversational command of general flow terminology.

B. Understand the subtleties of performance specification.

C. Know how flowmeters are calibrated.

A glossary of terms on any subject is at best a cumbersome list of foreign words that one must try to commit to memory in order to have a conversational command of the subject matter, and to understand more easily further reading or studies in that subject. Typically, a glossary is set up in the form of a list of words or phrases and their accompanying definitions. Learning terminology, when presented in list form, not only is boring but also mentally fatiguing, primarily because of the lack of visual aids to create a mental picture to associate with a word or definition.

Consequently, in this unit we will try to present a glossary of terms used in flow measurement in a more palatable manner by writing in a conversational form, highlighting the word or phrase to be remembered, and associating, where appropriate, the word with a visual aid. Let's try this technique on the terminology used to describe the physical parts of a flowmeter installation.

3-1. Flowmeter Physical Terms

A *flowmeter* is a device for measuring the quantity or rate of flow of a moving fluid in a pipe or open channel. It may consist of a *primary device* and a *secondary device*. This definition and those that follow are listed in an American National Standard (ANSI/ASME MFC-1M-1979 (Ref. 1).

Most flowmeters consist of two major parts called a *primary device* and a *secondary device*, as shown in the example presented in Fig. 3-1. The *primary device* generates a signal responding to the flow from which the flow rate may be inferred. In the case of the example head-class flowmeter shown in Fig. 3-1, the primary device is the orifice which infers the flow rate through its induced pressure drop.

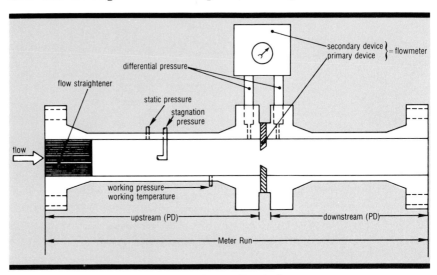

Fig. 3-1. Flowmeter Installation and Fluid-Related Terminology

A *secondary device* receives a signal from the primary device and displays, records, and/or transmits it as a measure of the flow rate. In the example, the differential-pressure transmitter is considered to be the *secondary device*. A good general rule to remember here is that, in most cases, the primary device is wetted by the fluid and the secondary device is outside the process.

A flowmeter, especially if high accuracy is required, is not normally installed in an existing pipeline between existing flanges. Rather, a *meter run* is designed. Often the flowmeter installed in the *meter run* is calibrated in a flow laboratory. A *meter run* is, as shown in Fig. 3-1, the upstream and downstream run of a straight length of pipe usually specified by a meter manufacturer or in a national or international standard, such as Ref. 2. Normally, the upstream and downstream lengths are expressed in terms of pipe diameter (PD), which is simply the length of pipe divided by the diameter of the pipe matching the flowmeter diameter.

Sometimes installed with the meter run are devices called *flow straighteners*, *profile regulators*, or *swirl removers*. A *flow straightener* is a general term used to describe any one of the variety of devices intended to reduce swirl and/or regulate the velocity profile. A *profile regulator* is a specific flow-straightening device inserted in a pipe to reduce the straight length required to achieve a fully developed velocity distribution. A *swirl remover* is a flow-straightening device inserted in a pipe to eliminate or reduce swirl, a rotation of the fluid around the centerline of the pipe commonly induced by elbows and reducers.

3-2. Fluid-Related Terms

When speaking of flow and flowmeters, the terms *flow rate, volume flow rate,* and *mass flow rate* will often be heard. There are important distinctions among these three terms which must be understood.

The term *flow rate* is most commonly used, and should be used, when referring to the actual velocity, V, of the fluid medium. Units of velocity most often are expressed as feet per second and meters per second in English and SI units, respectively. When the expression *volume flow rate*, Q is used, the area, A, of the full closed-conduit is employed along with the average velocity, V, to arrive at the total quantity of flow. The flow usually is expressed in units of gallons or cubic feet per minute, or cubic meters per minute.

$$Q = V \times A$$

The term *mass flow rate* applies to another, more explicit, way of expressing the quantity of fluid material delivered. *Mass flow rate* is the actual pounds or grams (mass) of material delivered per unit time. The quantity *mass flow rate*, $\dot{m}$, is simply the product of fluid density, ρ, full closed conduit area, A, and fluid velocity V.

$$\dot{m} = \rho \times A \times V$$

All the terms defined above are used when speaking of the fluid itself and are, therefore, included in this discussion on fluid-related terminology. When speaking of the actual flowmeter, two terms are used, *rate meter* and *quantity meter*. These terms, although considered flowmeter physical terms, are defined here to more emphatically define their differences. A *rate meter* is a flowmeter through which the fluid does not pass in isolated (separately counted) quantities but in a continuous stream. Integration with respect to time is necessary to determine the total quantity of fluid passed. The terms *flow rate*, *volume flow rate*, and *mass flow rate* are, therefore, commonly used when referring to *rate meters*.

On the other hand, a *quantity meter* is a flowmeter in which the flow is separated into known isolated quantities which are separately counted to determine the total volume passed through the meter. An independent measurement of time is necessary to determine flow rate. Remember, the type of meter referred to when speaking of the *quantity meter* is the positive-displacement meter which we have placed in the pulse class of the extractive energy approach to flow measurement. All other flowmeters are *rate meters*.

Since we have just introduced the fluid-related parameters of velocity and density, let's continue with the terminology and definition of the various forms of pressure and temperature.

There are several terms used to express pressure. Two simple, basic terms that require definition are *absolute pressure* and *gage pressure*. *Absolute pressure* is the combined local pressure induced by some source and the atmospheric pressure at the location of the measurement. Conversely, *gage pressure* is the difference between the local *absolute pressure* of the fluid and the atmospheric pressure at the place of the measurement.

When dealing with flow equations, we will always use *absolute pressure* in the calculations.

In order to define *stagnation pressure (total pressure)*, *static pressure*, and *dynamic pressure*, it is important that we first become familiar with one of the basic equations used in flow measurement, Bernoulli's equation. Bernoulli's equation is a form of the energy equation and states simply that

Stagnation pressure = static pressure + dynamic pressure

$$P_T = P + 1/2\rho V^2$$

This equation will be derived from the energy equation and developed further in Unit 4. The intent here is to define and explain the physical meaning of each term in the above equation.

Stagnation pressure is the pressure one would obtain if one could bring a flowing fluid to a standstill (to rest) isentropically (without any energy loss).

It is the pressure a person riding in a car feels if he or she puts a hand outside the window with the palm perpendicular to the flow direction. *Stagnation pressure* and *total pressure* are synonymous and are used interchangeably.

Static pressure is the pressure of a fluid that is independent of the kinetic energy of the fluid. The kinetic energy, being essentially the *dynamic pressure*, is associated with fluid motion; whereas, the static pressure is associated more with a fluid at rest, or the pressure one would measure while moving with the fluid. For example, a pressure tap, flush with the pipe wall, measures *static pressure*; and according to Bernoulli's equation, if the pipe flow velocity is increased, the *static pressure* would decrease as long as the *stagnation pressure* (total energy input) is kept constant.

Dynamic pressure is the increase in pressure above the *static pressure* that results from complete transformation of the kinetic energy of the fluid into potential energy. It is equal to $1/2 \rho V^2$, where ρ is the fluid density and V is the fluid velocity. The effect of this pressure is felt by an object placed in the flow, such as the hand held outside the car window.

Working pressure (flowing pressure) is the static pressure of the fluid immediately upstream of a primary device. Similarly, *working temperature* is the temperature of the fluid immediately upstream of a primary device.

Differential pressure is the static pressure difference generated by the primary device when there is no difference in elevation between the upstream and downstream pressure taps.

Head pressure is simply the expression of a pressure in terms of the height of fluid through the use of the following equation:

$$P = y\rho g$$

where ρ is the fluid density, y is the height of a column of fluid, and g is the acceleration of gravity.

Pressure loss is the irrecoverable lost pressure caused by the presence of a primary device in a pipe.

To help you understand where these pressures physically exist, the locations of the pressures defined above are indicated in the head-class flowmeter installation shown in Fig. 3-1.

In the world of fluid dynamics, it is often very convenient to deal with *nondimensional parameters* because the often exaggerated amplification of dimensional parameters caused by the units one is working with is eliminated. A *nondimensional parameter* is a ratio of two quantities expressed in the same engineering units. One such nondimensional parameter, useful when dealing with gas velocities especially, is the *Mach number*. The *Mach number* is the ratio of the fluid velocity to the velocity of sound in the fluid (both velocities must be expressed in the same engineering units) at the same temperature and pressure.

Other terms used when speaking of velocity or flow rate are *flow rate range* and *rangeability*. *Flow rate range* is the range of flow rates bounded by the minimum and maximum flow rates. *Rangeability* is the ratio of the maximum flow rate to the minimum flow rate of a meter.

Another very important dimensionless parameter used to define the regions within the operating flow rate range of a meter where the meter performance is sensitive or insensitive to fluid parameters is the *Reynolds number*. The *Reynolds number* is a ratio of the inertia and viscous forces in a fluid. It is defined by the formula

$$Re = \frac{\rho V l}{\mu} = \frac{\text{inertia}}{\text{viscosity}}$$

where ρ is the density, V is the velocity, μ is the viscosity, and l is a characteristic dimension of the system in which the flow occurs. When specifying *Reynolds number*, one should indicate the characteristic dimension on which it has been based; for example, diameter of the pipe, diameter of head class device, orifice or venturi throat, or diameter of a pitot tube head.

The *Reynolds number* is also used to describe the condition of the flow, whether it be *laminar, transitional,* or *turbulent*. *Laminar flow* is defined as flow under conditions where forces due to viscosity are more significant than forces due to inertia. This type of flow has the characteristic of having adjacent fluid particles move along essentially parallel paths. On the other hand, *turbulent flow* exists where forces due to inertia are more significant than forces due to viscosity. Here, adjacent fluid particles are more or less random in motion and do not move along parallel paths.

Transitional flow is flow between *laminar* and *turbulent flow*. For a transition flow, the *Reynolds number* referred to the pipe diameter is generally between a lower limit of 2000 and an upper limit which varies between 7000 and 12,000, depending on pipe roughness and other factors.

A physical representation of *laminar, transitional,* and *turbulent flows* and their corresponding velocity profiles is provided in Fig. 3-2.

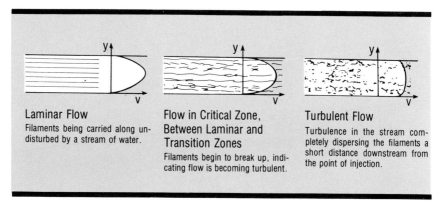

Fig. 3-2. Laminar, Transitional, and Turbulent Flows

In addition to the three flow regions defined above, there are three dynamic flow types encountered in practice: *steady flow, unsteady flow,* and *pulsating flow*.

Steady flow is a flow in which the flow rate in a measuring section does not vary significantly with time. The steady flows observed in pipes are, in practice, flows in which quantities (such as velocity, pressure, mass, density, and temperature) vary in time about mean values which are independent of time; these are actually "statistically steady flows."

Unsteady flow, on the other hand, is a flow in which the flow rate fluctuates randomly with time and for which the mean value is not constant.

Pulsating flow is defined as having a flow rate that varies with time, but for which the mean flow rate is constant when obtained over a sufficiently long period of time. Two types of *pulsating flow* are found in practice; periodic pulsating flow and fluctuating (random) pulsating flow.

3-3. Flowmeter Performance Terminology

One of the most difficult terms to define is the most important term in specifying flowmeter performance: *accuracy*. The qualitative definition of *accuracy* is the measure of freedom from error; the degree of conformity of the indicated value to the true value of the measured quantity. Essentially, *accuracy* has to do with the "closeness to the truth." However, what is the true value? How does one know this base from which errors are judged?

We will deal with these questions later in this unit; but first, we will discuss how *accuracy* is commonly specified by flowmeter manufacturers and what terms make up this elusive estimate of "truth" called *accuracy*.

There are two ways the accuracy of a flowmeter is specified: in *percent of span* or *percent of actual flow*. In Fig. 3-3, a graphic display of both accuracy specifications is provided. Note the ±1/2 *percent of actual flow* applies over the entire flow rate range as specified by the manufacturer, i.e., the device is accurate to ±1/2% of the flow rate whether the flow rate is 10, 20, 30, or 100 gallons per minute (GPM). On the other hand, the *percent-of-span* value of ±1/2 applies **only** at the maximum rated flow. To calculate the actual percent error over the entire flow rate range using the *percent-of-span* specification, the absolute error value obtained at full span is applied over the entire range. For example, using the data of Fig. 3-3, ±1/2% of 100 GPM (or full span flow) is ±0.5 GPM. Applying this value (±0.5 GPM) at 10 GPM results in an error of ±5% of actual flow.

$$\text{Percent of actual} = \frac{\pm 0.5 \text{ GPM} \times 100}{10 \text{ GPM}} = \pm 5\%$$

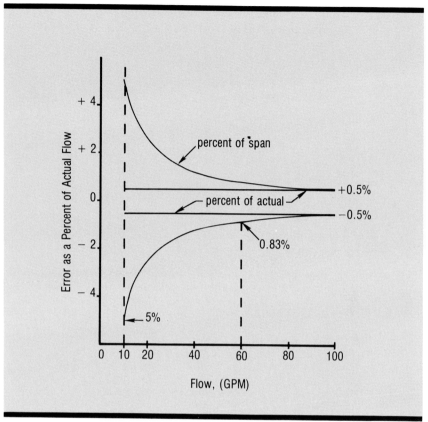

Fig. 3-3. Accuracy Envelope Boundary

Equivalently, at 60 GPM the actual error is ± 0.83%. Consequently, although a *percent-of-span* specification may be lower than a *percent-of-actual* specification, at lower flows considerable error allowances are provided for in the *percent-of-span* specification. So don't be fooled by high accuracy specifications when expressed in terms of *percent of span*.

What components make up what we have so far qualitatively defined as *accuracy* or the "closeness to the truth?"

A statement of accuracy may be expressed in terms of *uncertainty* and a *confidence level*. The interval within which the true value of a measured quantity is expected to lie, with a stated probability, is called *uncertainty*. The *confidence level* associated with the *uncertainty* indicates the probability that the interval quoted will include the true value of the quantity being measured. Consequently, an accuracy statement for a flowmeter may be, for example, ± 1% of actual flow rate *uncertainty at a 95% confidence level.* Simply stated, the manufacturer is confident that the instrument sold will be at least ± 1% from the "true" value 95% of the time. At present, confidence levels calculated using statistical methods usually are not given in the accuracy specification and accuracy is associated only with the uncertainty interval.

Of what does the uncertainty interval consist?

In very basic terms, the *uncertainty* interval consists of the *systematic error* and the *random error*. A graphic illustration of these two types of error is shown in Fig. 3-4. A *systematic error* is that which cannot be reduced by increasing the number of measurements if the equipment and conditions remain unchanged.

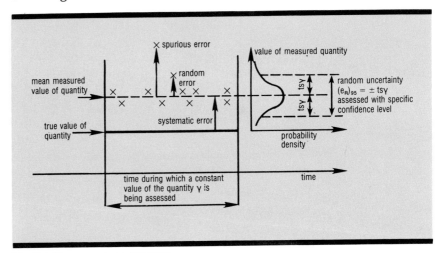

Fig. 3-4. Components of Accuracy

A *random error*, sometimes referred to as *precision* or *repeatability*, is caused by numerous, small, independent influences which prevent a measurement system from delivering the same reading when supplied with the same input value of the quantity being measured. In other words, the data points deviate from a mean value in accordance with the laws of chance.

Also shown in Fig. 3-4 is the *spurious error* or errors due to instrument malfunction or to human goof-ups. These errors should be caught by the manufacturer. They normally are not considered in the accuracy statement of the flowmeter. Therefore, the systematic error and the random error form the foundation for the accuracy specifications. However, a confidence level should be attached to the accuracy specification to complete the picture because the uncertainty and the confidence level are tied together statistically. For example, the uncertainty (accuracy) could be a very low number, indicating a high accuracy meter, if that number were calculated statistically at a very low confidence level. Therefore, the customer would only see the kind of accuracy he thought he bought a small percentage of the time he is using the meter.

Another term used to describe the performance of a flowmeter having linear output characteristics is *linearity*. Flowmeters having a zero-based linear frequency-velocity calibration characteristic, as shown in Fig. 3-5, are the turbine and vortex flowmeters. The performance of these flowmeters is usually expressed as *linearity*, which is the maximum percent deviation of the calibration data from a zero-based straight line. Linearity is calculated by first finding the mean slope which is simply the zero-based maximum slope $(f/V)_{max}$ added to the zero-based minimum slope $(f/V)_{min}$ and divided by two as shown in Fig. 3-5. The mean slope, also called the meter factor, is then the base from which the percent deviation for the remaining data is calculated.

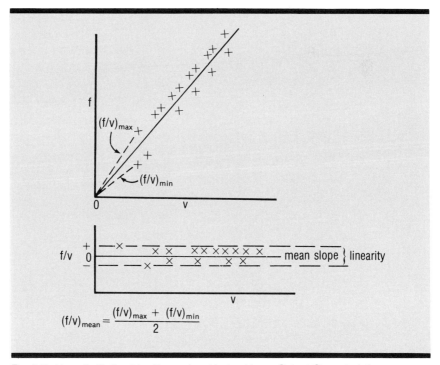

Fig. 3-5. Linearity Defined for Flowmeters Having Linear Output Characteristics

3-4. Flowmeter Calibration Terminology

The *calibration* of a flowmeter is the determination of the experimental relationship between the quantity being measured and the output of the device which measures it; where the quantity measured is obtained through a recognized standard of measurement. In flowmetering, there are two recognized standards of measurement; namely the *weighing method* (either *static weighing* or *dynamic weighing*) and the *volumetric method* (either *volumetric tank* or *prover*).

The *weighing method* is a standard method of measurement, suitable only for liquids where the flow is directed either intermittently (*static weight*) or continuously (*dynamic weight*) onto the scale of a weighing machine. Flow rate is obtained by measuring the mass of liquid accumulated in a known time interval.

Static weighing is a method in which the net mass of liquid collected is deduced from tare (empty tank) and gross (full tank) weighings respectively made before the flow is diverted into the weighing tank and after it is diverted to the by-pass.

In the *dynamic weighing* methods, the net mass of liquid collected is deduced from weigh readings taken while flow is directed into the weigh tank. A diverter is not used with this method. Instead, the weigh scale is preloaded so that when the weight of liquid collected reaches the pre-set value, the weigh beam moves to start a timer. A new, higher weight value is then preset so that when the weight of liquid collected reaches the higher value, the weigh beam moves to stop the timer. The two preset weights and the time are used to calculate the mass flow rate.

The *volumetric method* is a standard method of measurement in which the flow is directed into or out of a calibrated volumetric tank during a known period of time. A *volumetric tank* is a tank which will not lose its shape and the capacity of which has been determined by a primary method of calibration. *Proving* is a term used to define the determination of meter performance by establishing the relationship between the volume actually passed through the meter and the volume indicated by the meter.

Exercises

3-1. In head-class meters the **primary** devices are a variety of obstacles that restrict or impede flow through a pipe or channel. What type of **secondary** devices would you expect them to have?

3-2. A four-inch (pipe) pulse class meter allegedly registers 53.68 pulses per cubic foot of flow (this number is the **meter factor**). A proof test was run on this meter during which water was collected in a weigh tank. The following data were obtained:

Original tank weight (Tare Weight)	2008.2 lb
Final tank weight (Gross Weight)	3017.5 lb
Density of fluid (Water at 88.4 °F)	62.4 lb/cu ft
Elapsed time of test	211 sec
Number of observed pulses	907

(a) According to the meter, how many cubic feet of water ran through the pipe during the test?

(b) How much water (by weight) was actually collected in the tank during the test? How many cubic feet of water is this?

(c) Compute the percent error in the flow measurement observed by the meter using the relation

$$\text{Percent error} = \frac{(\text{observed value}) - (\text{actual value})}{(\text{actual value})} \times 100$$

(d) What is the actual volumetric flow **rate**? What flow **rate** is observed by the meter? Is the percent error in flow rate the same as the percent error in total flow computed in part (c)?

(e) What is the relationship between the **volumetric flow rate** (Q) and the mass **flow rate** ($\dot{m}$)?

(f) What is the technical name for this type of proof test?

3-3. Air at 20° C, viscosity ($\mu = 1.8 \times 10^{-5}$ kg/ms); Density ($\rho = 1.20$ kg/m³) is flowing at a one meter per second through a six-centimeter (.06 meter) diameter pipe.

(a) What is the **Reynolds number** for this flow configuration?

(b) Is the flow in the pipe **laminar, transitional** (partially **turbulent**), or fully **turbulent**?

(c) Would this ratio be different if viscosity was expressed in lb_m/ft sec, density in lb_m/ft³, velocity in feet per second, and the pipe diameter in feet?

3-4. A flowmeter is specified to have an **uncertainty** of 0.5% of **actual** with a **confidence level** of 95%.

(a) What percent of the meter measurements would you expect to be more than 0.5% high?

(b) Do these specifications assume the error to be **systematic** or **random** in nature?

3-5. Three flowmeters measure flows from 20 ft/sec to 100 ft/sec in a four-inch pipe. They are specified to perform as follows:

    Meter I: within ± 0.5% of **span**
    Meter II: within ± 0.5% of **actual**
    Meter III: within ± 0.5% **linearity**

(a) If each meter reads 100 ft/sec, what is the maximum possible error in this reading for each meter? What are the maximum and minimum possible values for the true flow rate through each meter?

(b) Repeat part (a) for meter readings of 60 ft/sec and 20 ft/sec.

(c) Do the specifications for each meter assume the error to be either **random** or **systematic** in nature?

(d) What are the **range** and **rangeability** of the three meters?

References

[1] "Glossary of Terms Used in the Measurement of Fluid Flow in Pipes," American National Standard ANSI/ASME MFC-1M-1979.

[2] "Measurement of Fluid Flow by Means of Orifice Plates, Nozzles, and Venturi Tubes Inserted in Circular Cross-section Conduits Running Full," International Standards Organization 5167-1980 (E).

Unit 4:
Theory of Differential Pressure Flowmetering Devices

Unit 4

Theory of Differential Pressure Flowmetering Devices

This unit is designed to teach the origin and development of both the laminar and the turbulent *head-class* flowmeter equations and the theory for both the liquid (incompressible) and the gas (compressible) flow.

Learning Objectives—When you have completed this unit, you should:

A. Understand the origin and development of the basic equation that describes all differential pressure flowmetering devices.

B. Know the differences between incompressible (liquid) and compressible (gas) flow measurement.

C. Understand the difference between the laminar and turbulent flow regimes.

This unit contains numerous equations and mathematical formulation. Please follow each step in the mathematical development because each step reveals how the final form is obtained. In addition, in the development of the final form, certain parameters are defined. It is important to understand the origin of these parameters because they are eventually used in the working equations for sizing head-class devices.

4-1. Incompressible Flow

The basic physical concept describing the flow of fluid in closed conduits is the conservation of energy. Energy cannot be created or destroyed. This concept is idealized in the basic flow equations, and the link to reality lies in the coefficient employed to correct for lost energy.

Consequently, let's start the development of the basic imcompressible flow equation from first considerations of energy conservation.

A flowing fluid, moving from "station 1" to "station 2" as shown in Fig. 4-1, may change in *potential* and *kinetic* energy; however, in the absence of friction its *total* energy remains unchanged. Because the fluid is incompressible, the mass m passing through any station of the stream tube in a given time is the same.

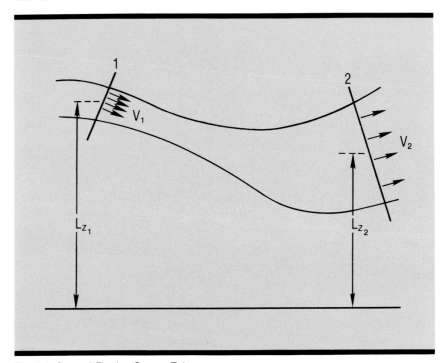

Fig. 4-1. General Flowing-Stream Tube

Expressed in terms of volume v, the mass m is

$$m = \rho v$$

where ρ is the fluid density.

The fluid mass passing station 1 is acted upon by an amount of work $P_1 v$ or $P_1 m/\rho$ and this same mass of fluid produces an equivalent amount of work at station 2, $P_2 m/\rho$ since there is no friction and the total energy remains the same. Because the net work $P_2 m/\rho - P_1 m/\rho$ produced by the mass m must equal the reduction in its potential and kinetic energy when moving from station 1 to station 2, the following mathematical form results:

$$(mgL_{z1} + 1/2 mV_1^2) - (mgL_{z2} + 1/2 mV_2^2) = \frac{P_2 m}{\rho} - \frac{P_1 m}{\rho}$$

Rearranging the terms by dividing both sides by mg, an expression in terms of unit weight evolves, namely:

$$L_{z_1} + \frac{V_1^2}{2g} + \frac{P_1}{\rho g} = L_{z_2} + \frac{V_2^2}{2g} + \frac{P_2}{\rho g}$$

where L_z = elevation head, $\frac{V^2}{2g}$ = velocity head, $\frac{P}{\rho g}$ = pressure head, and g = acceleration due to gravity.

The sum of these three terms is called the total head or stagnation pressure (P_T). Using this definition of stagnation pressure and considering a horizontal pipe line having the same elevation, $L_{z1} = L_{z2}$, the above equation reduces to the general form:

Total energy = potential energy + kinetic energy

$$P_T = P + 1/2\ \rho V^2$$

Stagnation pressure = static pressure + dynamic pressure

To summarize, Bernoulli's equation shows that for an incompressible friction-free flow, part of the stagnation pressure (which is the total pressure driving the flow) can be converted to dynamic pressure (which is related to kinetic energy), so that the actual (static) pressure at any point in the flow is related to the fluid velocity at that point. Thus, if the flow is constricted at some point (so that it must go **faster** through the constriction), the static pressure will be **lower** at that point. Most head-class meters measure the difference between working (upstream) pressure and the pressure at a constriction. The object of this section is to derive (from Bernoulli's law) an expression for the upstream velocity in terms of this pressure difference and the geometry of the constriction.

Conventional head-class meters in round pipes are devices which reduce the diameter of the flow stream. These devices (orifice plates, nozzles, and venturis) are often identified by a beta ratio, which is the ratio of the diameter of the constriction to the pipe diameter:

$$\beta = D_{const}/D_{pipe}$$

Other nonconventional head-class flowmeters often are identified by an equivalent beta ratio or by their percent reduction in flowstream area.

In deriving the head-class equation, we will look at two flowstream cross sections; one upstream of the constriction and one at the constriction (see Fig. 4-2).

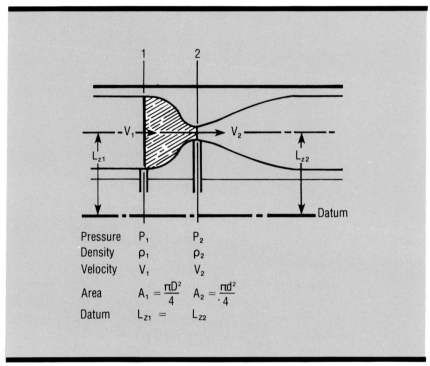

Fig. 4-2. Constriction in Horizontal Flow Line

If we now consider a constriction in a horizontal flow line, as shown in Fig. 4-2, the application of the conservation of energy concept results in the following equation:

$$P_1 + \frac{1}{2}\rho_1 V_1^2 = P_2 + \frac{1}{2}\rho_2 V_2^2$$

This equation states that the *total energy* (and *stagnation pressure*) at station 1 is equal to that at station 2. This can be restated in terms of *differential* pressure; for a fluid of constant density as:

$$\Delta P = P_1 - P_2 = \frac{1}{2}\rho(V_2^2 - V_1^2) \qquad (4\text{-}1)$$

The flow velocity at (V_2) can be found by applying the principle of conservation of mass. Because mass cannot be created or destroyed, the mass flowing into the meter must flow out of it, or

$$\dot{m}_1 = \dot{m}_2 = \rho A_1 V_1 = \rho A_2 V_2$$

and $$V_2 = \frac{A_1}{A_2} V_1$$

Substituting V_2 into Eq. (4-1) shows that

$$\Delta P = \frac{1}{2} \rho V_1^2 \left(\frac{A_1^2}{A_2^2} - 1 \right) \tag{4-2}$$

Since $A_1 = \frac{\pi D^2}{4}$ and $A_2 = \frac{\pi d^2}{4}$, Eq. (4-2) becomes

$$\Delta P = \frac{1}{2} \rho V_1^2 \left\{ \left(\frac{D}{d} \right)^4 - 1 \right\}$$

or $$V_1 = \sqrt{\frac{2\Delta P}{\rho}} \; \frac{1}{\sqrt{\left(\frac{D}{d}\right)^4 - 1}}$$

or $$V_1 = \sqrt{\frac{2\Delta P}{\rho}} \; \frac{1}{\sqrt{\frac{1}{\beta^4} - 1}} = \sqrt{\frac{2\Delta P}{\rho}} \; \frac{\beta^2}{\sqrt{1 - \beta^4}} \tag{4-3}$$

The term $\frac{1}{\sqrt{1 - \beta^4}}$ is referred to in the literature as the "velocity of approach" factor, denoted by the symbol "E." Thus, Eq. (4-3) becomes:

$$V_{1t} = \beta^2 E \sqrt{2\Delta P/\rho} \tag{4-4}$$

A "t" subscript has been introduced to show that this is the theoretical velocity.

The above development assumes that energy is conserved through a flowmeter. Actually, some energy is lost in the meter, due to friction losses. The actual pressure drop is larger than the theoretical value. Conversely, if the theoretical velocity (V_{1t}) is calculated using Eq. (4-4), the result will be higher than the actual velocity (V_{1a}). The ratio between these two numbers is a relatively constant factor called the discharge coefficient (C):

$$C = \frac{V_{1a}}{V_{1t}} = \frac{V_{1a}}{\beta^2 E \sqrt{2\Delta P/\rho}}$$

This coefficient is approximately 0.6 for orifice plates, 0.75 for nozzles, and 0.98 for venturi tubes.

The actual velocity, volume flow rate, and mass flow rate are now calculated as;

$$V_{1a} = \beta^2 CE \sqrt{2\Delta P/\rho} \qquad (4\text{-}5a)$$

$$Q_{1a} = A_1 V_{1a} = \frac{\pi^2}{4}(d^2\beta^2)CE\sqrt{2\Delta P/\rho} = \frac{\pi^2}{4}D^2 CE\sqrt{2\Delta P/\rho} \qquad (4\text{-}5b)$$

$$\dot{m}_{1a} = \rho(A_1 V_{1a}) = \rho\left[\frac{\pi^2}{4}D^2 CE\sqrt{2\Delta P/\rho}\right] = \frac{\pi^2}{4}D^2 CE\sqrt{2\rho\Delta P} \qquad (4\text{-}5c)$$

The discharge coefficient (C) and the velocity of approach factor (E) are both functions of β beta ratio. These numbers are always encountered together in the head class Eq. (4-5). For this reason, some engineers prefer to shorten these equations by introducing the flow coefficient (a), defined as

$$a = CE = C/\sqrt{1 - \beta^4}$$

Then Eq. (4-5) becomes;

$$V_{1a} = \beta^2 a \sqrt{\frac{2\Delta P}{\rho}} = \text{actual velocity} \qquad (4\text{-}6)$$

4-2. Compressible Flow

A gas passing through a constriction in a pipe behaves similarly to a liquid in that the pressure decreases as the velocity increases. The decrease in gas pressure, however, is accompanied by significant changes in density. A decrease in density is an indication of fewer molecules per unit volume, just as an increase is indicative of more molecules per unit volume. Consequently, gases are said to be compressible and the basic equations must allow for changes in density.

The derivation of the compressible form of Bernoulli's equation is laborious and beyond the scope of this learning module. Rather than develop the final form of the equations for compressible flow, it is more important to define the parameters of significance.

Consequently, the final form of the compressible-flow equations for mass and volume flow rate are the same as Eq. (4-5c) and Eq. (4-5b) except for one term called the "expansion factor." The expansion factor accounts for the change in density between stations 1 and 2 of the constricted flow depicted in Fig. 4-2. In terms of velocity, the expansion factor is defined as the ratio of the actual velocity to that predicted by the incompressible flow equation:

$$\varepsilon = \frac{V_{comp}}{V_{incomp}} = \frac{V_c}{\beta^2 CE \sqrt{2\Delta P/\rho_1}}$$

Note, the density, ρ, is that value measured at station 1.

The expansion factor in reality is a function of both the density at stations 1 and 2 as well as two parameters called the "isentropic exponent" and the "beta ratio." The isentropic exponent, γ, (gamma), is simply a ratio defined by the specific heat at constant pressure divided by the specific heat at constant volume. This value is dependent on the type of gas being considered. Certain gases have different values of the isentropic exponent. Presence of the isentropic exponent in the expansion factor arises from an assumption of no transfer of heat between the fluid and pipe. This implies no friction, permitting the assumption that any change of state between stations 1 and 2 is a reversible isentropic adiabatic change.

All of the above assumptions allow the calculation of the expansion factor, which has the final form:

$$\varepsilon = \left\{ \left(\frac{P_2}{P_1}\right)^{\frac{2}{\gamma}} \left(\frac{\gamma}{\gamma-1}\right) \left[\frac{1-(P_2/P_1)^{\frac{\gamma-1}{\gamma}}}{1-(P_2/P_1)}\right] \left(\frac{1-\beta^4}{1-\beta^4(P_2/P_1)^{\frac{2}{\gamma}}}\right) \right\}^{\frac{1}{2}}$$

The expansion factor is dependent on the diameter ratio, β, the pressure ratio, (P_2/P_1), and the ratio of specific heats, γ. Values for gas flowmeter-sizing calculations may be obtained from prepared curves or tables.

A thorough discussion on sizing head-class flowmeters and the use of the parameters defined above in the sizing calculations will be included in Units 5 and 6.

4-3. Laminar Flow

All real fluids possess a physical characteristic called "viscosity," which contributes to an increase in the internal friction when a fluid is in motion. Examples of particularly viscous liquids are honey, glycerine, and thick oils. When a fluid is highly viscous and the flow rate is low, the viscous forces outweigh the inertial forces, and the flow is described as being laminar. Because the viscous forces dominate, the basic flow equations for laminar flow are different from the turbulent flow equations.

The nature of viscosity is best portrayed by the following physical example. Consider the motion of a fluid held captive between two long parallel plates: one at rest, the other in motion at constant velocity as shown in Fig. 4-3. Experiments show that the fluid is attached to both walls so that the velocity at the lower plate is zero and the velocity at the upper plate is equal to the velocity of the plate. Further, the velocity distribution in the fluid between the plates is linear, and the fluid velocity is proportional to the distance, y, from the lower plate. The fact that the fluid has friction is demonstrated by the measurement of a force which resists the motion of the upper plate and has the magnitude:

$$\tau = -\mu \frac{V}{D}$$

per unit area of the plate.

A more rigorous statement may be made in that the shearing stress produced by the slip between layers of fluid is defined as:

$$\tau = -\mu \frac{dV}{dy}$$

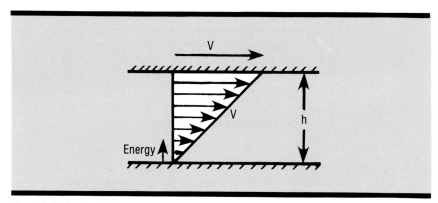

Fig. 4-3. Velocity Distribution in a Viscous Fluid between Two Parallel Flat Walls (Couette Flow)

where τ (shear stress) is the force per unit area in the flow direction acting on an element of area in a plane parallel to the y axis. The parameter μ is defined as the "coefficient of viscosity" and was described by Newton. Consequently, the above equation is often referred to as Newton's law of friction, and the fluids that conform to this law are considered to be Newtonian fluids.

The elementary law of friction for a simple flow may be applied to the more practical case of flow through a straight pipe of circular cross section. Consider laminar flow in a pipe of diameter D, where the velocity at the wall is zero and is a maximum on the axis, as in Fig. 4-4. At a long distance from the entrance section to the pipe, the velocity distribution across the pipe becomes independent of the pipe length. The fluid is moved by the pressure gradient acting in the flow direction, and, due to friction, individual layers act on each other with a shear stress proportional to the velocity gradient dV/dy. Consequently, a fluid particle is moved forward by the pressure gradient and retarded by internal friction.

Let's consider the condition of equilibrium for a fluid-filled pipe of length L and diameter D, as shown in Fig. 4-4. Equilibrium in the x direction requires the force $(P_1 = P_2)\, \pi y^2$ acting on the fluid to be equal to the shear force $2\pi y L \tau$ acting on the internal area of the pipe.

Therefore:

$$(P_1 - P_2)\, \pi y^2 = 2\, \pi y L \tau$$

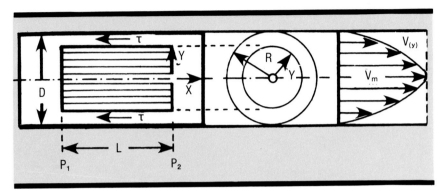

Fig. 4-4. Laminar Flow through a Pipe

or

$$\tau = \frac{(P_1 - P_2)}{L} \frac{y}{2} \tag{4-7}$$

If we now apply Newton's law of friction:

$$\tau = -\mu \frac{dV}{dy}$$

to Eq. (4-7) we obtain:

$$\frac{dV}{dy} = -\frac{(P_1 - P_2)}{\mu L} \frac{y}{2}$$

and upon integration we get:

$$V = \frac{(P_1 - P_2)}{\mu L} \left[C - \frac{y^2}{4} \right]$$

Because of the condition of no slip at the wall, $V=0$ at $y=\frac{D}{2}$. Therefore, the constant of integration becomes

$$C = \frac{D^2}{16}$$

and hence:

$$V = \frac{(P_1 - P_2)}{4\mu L} \left[\frac{D^2}{4} - y^2 \right]$$

Because the velocity distribution is parabolic, as seen in Fig. 4-4, and the y coordinate is zero at maximum velocity, the expression for maximum velocity becomes:

$$V_{max} = \frac{(P_1 - P_2)}{16\mu L} D^2$$

In terms of volumetric flow Q, the above equation becomes

$$Q = \frac{\pi D^2}{8} V_{max} = \frac{\pi D^4}{128 \mu L} (P_1 - P_2) \qquad (4\text{-}8)$$

because the volume of a paraboloid of revolution is equal to 1/2 x area x height.

As seen above, the volumetric flow rate is proportional to the first power of the pressure drop per unit length of pipe and the fourth power of the pipe diameter. If we introduce the expression for the mean velocity over the cross section as:

$$V_{mean} = \frac{Q}{A} = \frac{4Q}{\pi D^2}$$

Then Eq. (4-8) becomes:

$$V_{mean} = \frac{D^2(P_1 - P_2)}{32 \mu L} \qquad (4\text{-}9)$$

the final form for the mean velocity in a pipe whose flow is in the laminar regime.

The final form of the laminar equation was experimentally proven by G. Hagen in 1839 and discovered independently by Poiseuille. Consequently, the final equation is often referred to as the Hagen-Poiseuille law.

4-4. Turbulent Flow

Types of flow to which Eq. (4-8) and Eq. (4-9) apply exist in small-radii pipe at extremely low velocities. At higher velocities and larger radii, the pressure drop in a pipe ceases to be proportional to the first power of the mean velocity and becomes proportional to the second power of mean velocity.

The general expression for pressure drop in a straight pipe for turbulent flow is known as Darcy's formula, and has the form:

$$\Delta P = \frac{\rho f L V^2_{mean}}{2D} \qquad (4\text{-}10)$$

which is a form of Bernoulli's equation with the addition of a friction factor, f, pipe length, L, and diameter, D. The friction factor for laminar flow conditions ($R_D < 2000$) is a function of Reynolds number only, whereas for turbulent flow ($R_D > 8000$) it is a function of the pipe wall roughness. Widely accepted data of friction factors for use with the Darcy equation are presented in Refs. 1 and 2.

Exercises

4-1. A 4.00-inch diameter pipe contains an orifice-plate flow meter. The orifice plate has a beta ratio of 0.60. A pressure drop of 5.00 psi (lb_f/in^2) is observed when water at a density of 62.4 lb_m/ft^3 is run through the meter.

(a) Calculate the diameter of the orifice in the orifice plate.

(b) Calculate the velocity of approach factor (E) of this orifice plate.

(c) Calculate the theoretical velocity (V_t) of the water in the pipe.

(d) If the discharge coefficient for this pipe size, fluid, and β ratio is $C = 0.615$, calculate the actual velocity (V_a).

(e) Calculate the actual volume flow and mass flow rates.

4-2. Air, at a density of 0.075 lb_m/ft^3, is run through a flow meter identical to that in Exercise 4-1. Again, a 5 psi pressure drop is observed across the meter.

(a) (Optional) If the isentropic exponent for air is $\gamma = 1.4$ and the upstream pressure is 50 psi, show that the expansion factor (ε) for this flow configuration is 0.935.

(b) Given that the discharge coefficient for this flow configuration is $C = 0.615$, calculate the actual velocity, volume flow rate, and mass flow rate.

4-3. (a) Calculate the Reynolds number of the flow in Exercise 4-1 for the viscosity of water, $\mu = 6.7 \times 10^{-4} lb_m/ft\ sec$, using the pipe diameter as the reference length. Is this flow laminar or turbulent?

(b) Assuming the flow is through a four-inch cast iron pipe whose friction factor is approximately 0.027, and the meter is mounted in a 2000-foot section of pipe, calculate the pressure needed to drive the system.

(c) Water is run through 2000 feet of one-inch diameter pipe at 0.2 ft/sec. What is the Reynolds number of this flow? Is it laminar or turbulent? Find the pressure needed to drive the system.

References

[1] Moody, L. F., "Friction Factors for Pipe Flow," *Transactions of the American Society of Mechanical Engineers*, Volume 66, November 1944, pages 671 to 678.
[2] "Flow of Fluids through Valves, Fittings, and Pipe," Technical Paper No. 410, pages A-23 to A-25, Crane Co., 4100 S. Kedzie Avenue, Chicago, Illinois 60632.

Unit 5: Head-Producing Flowmeters I (Conventional)

Unit 5

Head-Producing Flowmeters I (Conventional)

Flow-measurement techniques that use the differential pressure generated by a primary device are quite flexible and may be made to suit the conditions of many installations. All the factors necessary to determine which primary device to use are virtually impossible to outline. However, in this unit the advantages and disadvantages of the three most commonly used primary devices are discussed to provide the student with an outline of their general field of use.

Learning Objectives—When you have completed this unit you should:

    A. Understand the advantages and disadvantages of the three conventional head-class meters discussed.

    B. Know how to size an orifice, venturi, and flow nozzle for a simple application.

There are many ancient examples applying the head-meter principle; namely, the hourglass and the use of a constriction or orifice to measure water flow to householders in Rome during Ceasar's time.

About the start of the seventeenth century, it was recognized that the volume rate of flow was equal to the velocity times the area and the flow rate through an orifice varied as the square root of the "head," differential pressure. In 1738 John Bernoulli developed the basic equation describing the flow through a head-type flowmeter and in 1797 Giovanni Battista Venturi published the results of his work on the venturi tube.

Clemens Herschel, using Venturi's basic work, developed the commercial venturi tube in 1887, Ref. 1, which proved to be a practical and economical means of measuring large volumes of flowing liquids. Measurement of gas flow using the venturi tube proved to be too complicated and impractical to apply to the commercial measurement of gas, because of a lack of understanding of the correction factors involved. These roadblocks and the need for a practical means for measuring large volumes of natural gas in the early 1900s led to investigation of the sharp, square-edged, thin-plate orifice, Ref. 2, which not only proved far more satisfactory for gas flow applications, but better in many other commercial applications.

First instance of the commercial use of the flow nozzle for the measurement of flow rate is difficult to trace. Reference to investigations of the principles of the nozzle extend back into the nineteenth century, Ref. 3.

Let's consider first the characteristics and application of the orifice plate.

5-1. Orifice Plates

There are basically four orifice plate types or designs: the *concentric*, which is most common, the *eccentric*, the *segmental*, and the *quadrant-edge*. A sketch of the four types is provided in Fig. 5-1. In addition to the different types of orifice plates, there are three different pressure tap locations commonly used: *flange taps, D and D/2 taps,* and *corner taps* (see Fig. 5-2 for a visual description of the pressure-tap locations).

In general, the orifice plate has the inherent advantage of being easy and inexpensive to replace, however, the initial installation may be costly due to the requirement of special orifice-plate flanges containing pressure taps. A great advantage over other types of flowmeters is that the orifice-plate primary device has no moving parts and the differential pressure sensor can be removed and replaced, if faulty, without shutting down the process. Of course, these advantages apply also to the nozzle and venturi.

The thin-plate, concentric orifice is most commonly used to measure the steady flow of clean homogeneous fluid, liquid, or gas in the turbulent flow regime and the greatest body of data and experience is available for the concentric plate. In the United States, flange taps are most commonly used with the concentric orifice, however, in Europe corner taps are favored. The eccentric plate has a bore (hole) which is tangent to the inside wall of the pipe and the hole is set either at the bottom of the pipe to allow for easy passage of nonabrasive solids (usually not more than 5%) or at the top of the pipe to allow passage of small amounts of gas in a liquid flow. Discharge coefficients for eccentric plates are not as predictable as those for concentric plates and errors three to five times greater than for the concentric plate can be experienced. Segmental orifice plates are also used in light slurries or exceptionally dirty gases. However, it is not as well accepted as the eccentric plate because of a scarcity of data and published experiences. The quadrant-edge orifice plate has a concentric opening, however, the upstream edge is round rather than sharp as in the case of the concentric orifice. This type of plate produces a discharge coefficient that remains constant in the low Reynolds number region normally experienced with the flow of viscous liquids.

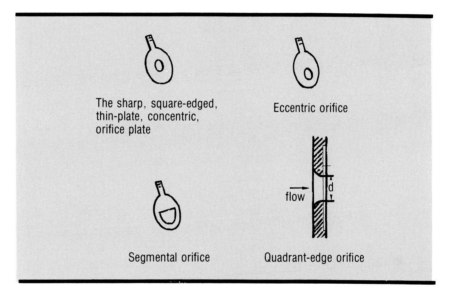

Fig. 5-1. Types of Orifice Plates

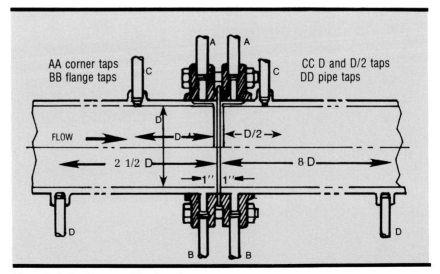

Fig. 5-2. Orifice Pressure Tap Locations

Sizing of an orifice plate to satisfy a particular process requirement is a fairly straightforward operation. Although semi-empirical, the sizing calculations are based on the Bernoulli equation derived from basic balance of energy considerations, as shown in the previous unit.

If we now refer to Eq. (4-5a) of Unit 4, it is seen that five basic parameters are present.

$$C = \frac{V\sqrt{1-\beta^4}}{\beta^2 \sqrt{\frac{2\Delta P}{\rho}}} \tag{5-1}$$

Of the five basic parameters, only three nomally are known or are specified by the process engineer. For example, the flow rate, V, velocity and ρ, fluid density, are normally known because the engineer knows the process flow-rate range and process fluid parameters. Further, the engineer has the freedom to specify the ΔP, differential-pressure range he would like to work with. Some companies may prefer to use only one or two differential-pressure range transmitters to affect commonality and thus, an interchangeability of parts. In this case, the engineer must use a transmitter with a prescribed maximum ΔP, differential-pressure range.

Consequently, the "beta ratio," β, and discharge coefficient, C, are the only unknowns in the equation and the sizing calculation is reduced to the determination of these unknowns so that Eq. (5-1) is satisfied over the flow rate and differential-pressure range prescribed.

To demonstrate how the unknowns are determined, a sample problem is provided in which each step in the sizing procedure is outlined in detail.

Consider sizing a sharp, square-edged, thin-plate concentric orifice for the following process conditions:

1. Pipe line: four-inch schedule 40 carbon steel
2. Process fluid: water

3. Process temperature: 70°F
4. Maxium flow: Q_{max} = 200 gpm, or V_{max} = 5.04 ft/sec
5. Maximum ΔP: ΔP_{max} = 100 inches H_2O, or ΔP_{max} = 520.4 lb_f/ft^2

Step 1. Knowing the process fluid is saturated water, and knowing the process temperature, find the density, ρ, from Table 5-1 as 62.30 lb_m/ft^3 or 1.935 slugs/ft^3.

Temperature °F	Density lb_m/ft^3	Absolute Viscosity lb_m/ft sec	Kinematic Viscosity* ft^2/sec
40	62.43	0.0010382	0.000016631
45	62.42	0.0009531	0.000015269
50	62.41	0.0008784	0.000014074
55	62.39	0.0008124	0.000013021
60	62.37	0.0007539	0.000012088
65	62.34	0.0007018	0.000011259
70	62.30	0.0006554	0.000010519
75	62.26	0.0006138	0.000009858
80	62.22	0.0005763	0.000009263
85	62.17	0.0005423	0.000008724
90	62.11	0.0005115	0.000008235
95	62.05	0.0004834	0.000007790
100	61.99	0.0004577	0.000007383
105	61.93	0.0004342	0.000007012
110	61.86	0.0004126	0.000006670
115	61.79	0.0003927	0.000006356
120	61.71	0.0003744	0.000006067
125	61.63	0.0003574	0.000005799
130	61.55	0.0003417	0.000005551
135	61.47	0.0003271	0.000005321
140	61.38	0.0003135	0.000005108
145	61.29	0.0003008	0.000004908
150	61.19	0.0002890	0.000004722
155	61.10	0.0002779	0.000004549
160	61.00	0.0002675	0.000004386
165	60.90	0.0002578	0.000004233
170	60.79	0.0002486	0.000004090
175	60.69	0.0000240	0.000003955
180	60.58	0.0002319	0.000003828
185	60.47	0.0002242	0.000003709
190	60.35	0.0002170	0.000003595
195	60.24	0.0002102	0.000003489
200	60.12	0.0002037	0.000003388

*Kinematic viscosity = (absolute viscosity) ÷ (density)

Table 5-1. Density and Viscosity of Liquid Water at 1 atm.

Step 2. Using the appropriate units, substitute values of $V_{max} = 5.04$ ft/sec, $\Delta P_{max} = 520.4$ lb$_f$/ft² and $\rho = 1.935$ slugs/ft³ in Eq. (5-1) to arrive at:

$$C = 0.2173 \frac{\sqrt{1-\beta^4}}{\beta^2}$$

Step 3. To estimate a β value that satisfies the pressure differential requirements, a nominal value of 0.6 is always chosen for the discharge coefficient, C. Consequently, Eq. (5-1) reduces to:

$$2.76 \beta^2 = \sqrt{1 - \beta^4}$$

and solving for β results in:

$$\beta = 0.5837$$

To insure that the maximum desired ΔP of 100 inches H$_2$O is not exceeded, round off the β ratio to the next second decimal place so that the orifice opening is larger and a slightly lower differential pressure will result.

Therefore: $\beta = 0.59$ (always round off to the higher value).

Step 4. Once the β ratio is chosen, the discharge coefficient may be calculated. Before we discuss the discharge coefficient calculation procedures, both the bore and pipe Reynolds numbers are needed, and before the Reynolds numbers are calculated the fluid viscosity must be obtained. From Table 5-1 for water at 70°F obtain a viscosity, μ, of 6.55×10^{-4} lb$_m$/sec-ft.

Step 5. Now calculate the bore Reynolds number based on d, the orifice bore diameter and the pipe Reynolds number based on D, the pipe diameter.

$$R_d = \frac{\rho V_{bore} d}{\mu} = \frac{\rho V D}{\mu \beta} = \frac{62.30 \times 5.04 \times 4.026}{0.655 \times 10^{-3} \times 12 \times 0.59}$$

$$R_d = 272{,}595$$

$$R_D = \frac{\rho V D}{\mu} = \frac{62.30 \times 5.04 \times 4.026}{0.655 \times 10^{-3} \times 12}$$

$$Re_D = 160{,}831$$

At this point we are ready to calculate the exact value of the discharge coefficient, C, using the estimated value of β, the Reynolds number, and specifying the type of pressure taps to be employed. There are two recognized methods used to calculate the discharge-coefficient value. However, flow-laboratory calibration, although costly, is the method producing the greatest accuracy. The two recognized calculation methods are the ASME (American Society of Mechanical Engineers) equations as presented in Ref. 4, *Fluid Meters* and the Stolz equation as presented in Ref. 5, ISO (International Standards Organization) 5167. Both equations evolved from curve fitting many data sets obtained on orifice plates having a wide range of β values operating over a wide Reynolds number range. Because these equations are based on data, the sizing method is considered semi-empirical in nature.

The ASME equation, being a bit more complicated than the Stolz equation, will not be reproduced here. However, we will use the ASME tables of discharge coefficient.

Step 6. Choosing *flange taps* as the process interface mechanism, enter Table 5-2 listing the ASME discharge coefficients for a four-inch pipe. For a beta ratio of 0.59 and a bore Reynolds number $R_d = 272{,}595$, note immediately the tabulated discharge coefficients are to the right of the solid black line separating the coefficients. The uncertainty on the coefficients above and to the right of the black line do not exceed ±1%, whereas coefficients below and to the left of the line are extrapolations outside the range of the test data used and are subject to larger uncertainties. Consequently, it is important to be operating in the range where the equation represents actual data.

From a double interpolation of the data in Table 5-2 using $\beta = 0.59$ and $R_d = 272{,}595$, obtain a value of discharge coefficient

$$C = 0.6092$$

The alternate approach would be to use the full set of ASME equations presented in Ref. 4, *Fluid Meters*.

β \ R_d	10,000	12,000	14,000	16,000	18,000	20,000	25,000	30,000	40,000	50,000	75,000	100,000	500,000	1,000,000
.1500	.6109	.6089	.6075	.6065	.6056	.6050	.6038	.6030	.6020	.6014	.6006	.6002	.5993	.5992
.2000	.6098	.6077	.6061	.6050	.6041	.6034	.6021	.6012	.6001	.5995	.5986	.5982	.5972	.5970
.2500	.6104	.6081	.6064	.6052	.6043	.6035	.6021	.6012	.6001	.5994	.5985	.5980	.5969	.5968
.3000	.6125	.6100	.6083	.6070	.6059	.6051	.6037	.6027	.6015	.6007	.5998	.5993	.5981	.5979
.3500	.6156	.6130	.6110	.6096	.6085	.6076	.6060	.6049	.6036	.6028	.6017	.6012	.5999	.5997
.4000	.6197	.6166	.6145	.6128	.6115	.6105	.6087	.6075	.6059	.6050	.6038	.6032	.6017	.6015
.4500	.6249	.6213	.6187	.6168	.6153	.6140	.6119	.6104	.6086	.6075	.6061	.6053	.6036	.6034
.5000	.6314	.6270	.6238	.6215	.6196	.6182	.6155	.6138	.6116	.6102	.6085	.6076	.6055	.6052
.5500	.6391	.6337	.6298	.6269	.6246	.6228	.6195	.6174	.6147	.6130	.6109	.6098	.6072	.6068
.5750	.6434	.6374	.6331	.6298	.6273	.6253	.6217	.6193	.6163	.6145	.6121	.6109	.6080	.6076
.6000	.6479	.6412	.6365	.6329	.6301	.6279	.6239	.6212	.6179	.6158	.6132	.6118	.6086	.6082
.6250	.6525	.6451	.6399	.6359	.6328	.6304	.6259	.6230	.6193	.6171	.6141	.6126	.6091	.6087
.6500	.6571	.6489	.6431	.6388	.6354	.6327	.6278	.6245	.6205	.6180	.6148	.6131	.6092	.6087
.6750	.6614	.6525	.6461	.6413	.6376	.6346	.6292	.6256	.6212	.6185	.6149	.6131	.6088	.6083
.7000	.6652	.6554	.6484	.6432	.6391	.6359	.6300	.6261	.6218	.6183	.6144	.6124	.6077	.6071
.7250	.6697	.6591	.6515	.6458	.6413	.6378	.6314	.6271	.6218	.6186	.6143	.6122	.6071	.6065
.7500	.6788	.6672	.6589	.6526	.6478	.6439	.6369	.6323	.6265	.6230	.6183	.6160	.6104	.6097

Table 5-2(a). Flange Taps: Discharge Coefficients, C, for Square-Edged Orifices (Two-inch pipe, D = 2.067 in)

β \ R_d	10,000	12,000	14,000	16,000	18,000	20,000	25,000	30,000	40,000	50,000	75,000	100,000	500,000	1,000,000
.1500	.6126	.6094	.6071	.6054	.6041	.6030	.6011	.5998	.5982	.5973	.5960	.5954	.5939	.5937
.2000	.6147	.6114	.6090	.6072	.6058	.6047	.6026	.6013	.5996	.5986	.5973	.5966	.5950	.5948
.2500	.6167	.6133	.6108	.6090	.6076	.6064	.6044	.6030	.6013	.6003	.5989	.5983	.5966	.5964
.3000	.6187	.6152	.6127	.6108	.6093	.6082	.6061	.6047	.6029	.6019	.6005	.5998	.5981	.5979
.3500		.6178	.6152	.6132	.6116	.6103	.6081	.6066	.6047	.6036	.6021	.6013	.5995	.5993
.4000		.6219	.6188	.6166	.6148	.6133	.6108	.6091	.6069	.6056	.6039	.6031	.6010	.6008
.4500		.6278	.6241	.6213	.6192	.6175	.6144	.6123	.6097	.6082	.6061	.6051	.6026	.6023
.5000			.6311	.6276	.6249	.6228	.6189	.6163	.6131	.6111	.6086	.6073	.6042	.6038
.5500			.6398	.6354	.6320	.6292	.6243	.6210	.6169	.6144	.6112	.6095	.6056	.6051
.5750			.6448	.6398	.6359	.6329	.6273	.6236	.6190	.6162	.6125	.6107	.6062	.6057
.6000			.6501	.6445	.6401	.6367	.6304	.6263	.6211	.6179	.6138	.6117	.6067	.6061
.6250				.6493	.6445	.6406	.6336	.6289	.6231	.6196	.6149	.6126	.6069	.6063
.6500				.6541	.6487	.6444	.6366	.6313	.6248	.6209	.6157	.6131	.6070	.6061
.6750				.6587	.6527	.6479	.6392	.6334	.6262	.6219	.6161	.6132	.6062	.6054
.7000				.6628	.6562	.6508	.6413	.6349	.6269	.6221	.6157	.6125	.6049	.6039
.7250					.6594	.6535	.6430	.6360	.6272	.6220	.6150	.6115	.6031	.6020
.7500					.6634	.6571	.6456	.6379	.6283	.6226	.6149	.6111	.6019	.6007

Table 5-2(b). Flange Taps: Discharge Coefficients, C, for Square-Edged Orifices (Four-inch pipe, D = 4.026 in)

Table 5-2(c). Flange Taps: Discharge Coefficients, C, for Square-Edged Orifices (Eight-inch pipe, D = 7.981 in)

β \ R_d	14,000	16,000	18,000	20,000	25,000	30,000	40,000	50,000	75,000	100,000	500,000	1,000,000
.1500	.6166	.6137	.6114	.6096	.6064	.6042	.6015	.5999	.5977	.5967	.5941	.5937
.2000	.6184	.6155	.6132	.6114	.6081	.6060	.6032	.6016	.5994	.5983	.5957	.5954
.2500	.6190	.6162	.6140	.6122	.6091	.6070	.6044	.6028	.6007	.5997	.5971	.5968
.3000	.6197	.6170	.6148	.6131	.6101	.6080	.6055	.6039	.6019	.6009	.5984	.5981
.3500	.6217	.6189	.6167	.6149	.6117	.6096	.6070	.6054	.6032	.6022	.5996	.5993
.4000	.6261	.6228	.6203	.6183	.6147	.6123	.6093	.6074	.6050	.6038	.6009	.6006
.4500	.6334	.6294	.6263	.6238	.6193	.6164	.6126	.6104	.6074	.6059	.6023	.6019
.5000	.6443	.6390	.6350	.6318	.6259	.6220	.6172	.6142	.6104	.6084	.6037	.6032
.5500	.6586	.6518	.6464	.6421	.6344	.6292	.6228	.6190	.6142	.6112	.6051	.6043
.6000		.6592	.6531	.6482	.6393	.6334	.6260	.6216	.6157	.6127	.6056	.6047
.6250		.6674	.6603	.6547	.6446	.6378	.6294	.6243	.6175	.6142	.6061	.6050
.6500			.6680	.6616	.6501	.6424	.6328	.6270	.6193	.6155	.6062	.6051
.6750			.6759	.6687	.6556	.6469	.6361	.6295	.6208	.6165	.6061	.6048
.7000				.6757	.6610	.6513	.6391	.6317	.6220	.6171	.6053	.6039
.7250				.6823	.6660	.6551	.6415	.6333	.6220	.6170	.6039	.6023
.7500					.6705	.6584	.6433	.6343	.6222	.6162	.6017	.5999
					.6749	.6617	.6451	.6351	.6219	.6153	.5993	.5973

Table 5-2(d). Flange Taps: Discharge Coefficients, C, for Square-Edged Orifices (Sixteen-inch pipe, D = 15.25 in)

β \ R_d	18,000	20,000	25,000	30,000	40,000	50,000	75,000	100,000	500,000	1,000,000
.1500	.6244	.6213	.6158	.6122	.6076	.6049	.6012	.5994	.5950	.5944
.2000	.6249	.6220	.6167	.6131	.6087	.6060	.6025	.6007	.5965	.5960
.2500	.6236	.6210	.6161	.6129	.6089	.6065	.6032	.6016	.5977	.5973
.3000			.6157	.6128	.6091	.6068	.6039	.6024	.5989	.5984
.3500			.6167	.6138	.6101	.6079	.6050	.6035	.5999	.5995
.4000			.6202	.6168	.6127	.6102	.6068	.6052	.6011	.6006
.4500			.6269	.6226	.6172	.6140	.6098	.6076	.6025	.6018
.5000				.6314	.6241	.6197	.6139	.6110	.6039	.6031
.5500				.6433	.6332	.6272	.6191	.6150	.6054	.6042
.5750				.6504	.6386	.6315	.6221	.6173	.6060	.6046
.6000				.6581	.6444	.6362	.6252	.6197	.6065	.6049
.6250				.6664	.6505	.6410	.6283	.6220	.6068	.6049
.6500				.6749	.6568	.6459	.6314	.6241	.6067	.6045
.6750				.6836	.6630	.6506	.6341	.6259	.6061	.6036
.7000				.6919	.6687	.6548	.6363	.6270	.6047	.6020
.7250				.6999	.6740	.6585	.6378	.6275	.6026	.5995
.7500				.7077	.6791	.6619	.6390	.6276	.6001	.5966

A relatively simple equation, the Stolz equation is the second method one may choose to calculate the discharge coefficient.

The discharge coefficient C is given by:

$$C = 0.5959 + 0.0312\, \beta^{2.1} - 0.184\, \beta^8 + 0.0029\, \beta^{2.5} \left[\frac{10^6}{R_D}\right]^{0.75}$$

$$+\ 0.09\, L_1 \beta^4 (1-\beta^4)^{-1} - 0.0337\, L'_2 \beta^3$$

Note: When $L_1 \geq \dfrac{0.039}{0.09} = 0.4333$, use 0.039 for coefficient of $\beta^4 (1-\beta^4)^{-1}$ term.

In the equation:
- C = discharge coefficient
- β = diameter ratio d/D
- R_D = pipe Reynolds number, which must be between 10^4 and 10^7
- L_1 = distance of upstream pressure tap from the upstream face of the plate, divided by the pipe diameter
- L'_2 = distance of downstream pressure tap from *downstream* face of the plate, divided by the pipe diameter

For corner taps:

$$L_1 = L'_2 = 0$$

For D and D/2 taps:

$$L_1 = 1$$
$$L'_2 = 0.47$$

For flange-pressure taps:

$$L_1 = L'_2 = 1/D \text{ (D in inches)}$$

Now substituting $\beta = 0.59$, $R_D = 160{,}831$ and $L_1 = L'_2 = 0.2484$

for flange taps, into the Stolz equation obtain:

$$C = 0.6079$$

There is a 0.2% difference between the value obtained from the Stolz equation and the value obtained from the ASME tables. Either value is acceptable.

Step 7. Using the lower value of C = 0.6079, now check the maximum ΔP expected at the maximum velocity, V, using a 0.59 β ratio orifice.

$$0.6079 = \frac{5.04 \sqrt{1 - (0.59)^4}}{(0.59)^2 \sqrt{\frac{2 \times \Delta P}{1.935}}}$$

or ΔP_{max} = 482.33 $\frac{lb_f}{ft^2}$ or 92.68 in. H_2O

Therefore, a 100 in. H_2O differential pressure cell will be sufficient to cover the flow range desired.

An important rule-of-thumb to remember about head class devices, in general, is that the flow range over which the meter is accurate within the uncertainty specified on the discharge coefficient is approximately three to one. Therefore, in our sample case, if V = 5.04 ft/sec is the maximum flow rate of interest, the lowest flow rate measured with accuracy is on the order of 1-1/2 ft/sec or a measured differential pressure of 42.7 lb_f/ft^2 or 8.2 inches of water.

Step 8. The last step in the sizing procedure is to check the installation requirements of upstream length of pipe. To accomplish this, enter the appropriate graph of Fig. 5-3 with the known β ratio and note, for example, eight diameters of straight pipe are required after a single elbow. If proper installation practices are not followed, the discharge coefficient values could be quite different from the calculated values.

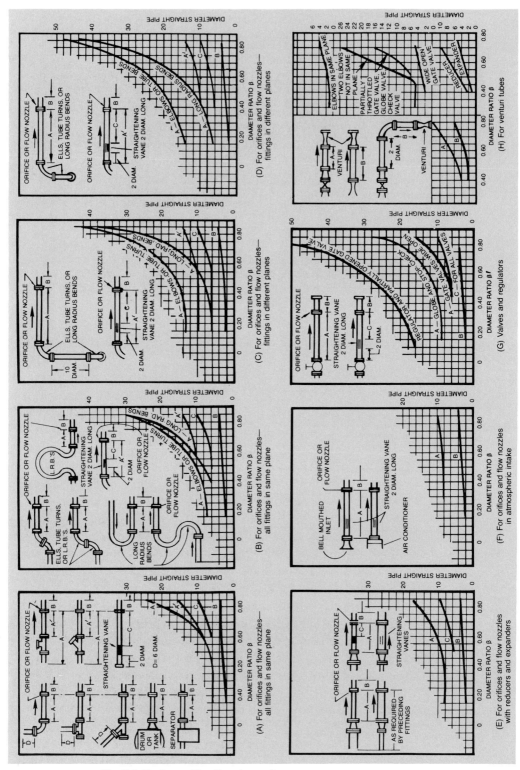

Fig. 5-3. Upstream Pipe Requirements

5-2. Venturi Tubes

There are basically two types of venturi tubes in use today: the classical or Herschel venturi tube and the venturi nozzle. As with the orifice plate, European preferences are different from those in the United States. The classical venturi is often used in the U.S.A., whereas the venturi nozzle is preferred in Europe. A sketch of both tubes is shown in Fig. 5-4. Note, the venturi nozzle is shorter than the classical tube. A shorter length is a decided advantage and the required length of the venturi tube is its biggest disadvantage. Long-length flowmeters are more costly, more difficult to install, and generally difficult to handle and store.

Advantages of the venturi over the orifice plate are its capacity to handle more flow while imposing less permanent pressure loss on the system. For example, for the same β ratio, a venturi is capable of 60% greater flow capacity than the orifice and the pressure loss is only 10% to 20% of the measured differential pressure. Other advantages are its ability to be used with fluids containing a relatively high percentage of entrained solids and its greater accuracy over a wider flow-rate range.

Three groups of classical venturi tubes are recognized by international standards organizations: the *rough cast* or *unmachined* converging cone, the *machined* converging cone, and the *rough-welded sheet metal* converging cone. Rough-cast tubes are used in pipe lines four inches to 32 inches in diameter, whereas machined tubes are seldom used in pipes larger than 10 inches in diameter. Rough-welded sheet-metal tubes are sometimes as large as 10 feet in diameter.

Recommended venturi tube proportions are shown in Fig. 5-5. The inlet section is a converging cone having an included angle of 21 degrees. This section is joined by a smooth curve to a short cylindrical section called the venturi *throat*. Another smooth curve joins the throat to a diffuser; a seven-to-eight degree included angle cone, which serves to recover a good deal of the pressure normally lost in the orifice-type device.

Four or more pressure taps are recommended in the inlet and throat sections. These taps should lead to a manifold so that an average pressure is obtained at each section.

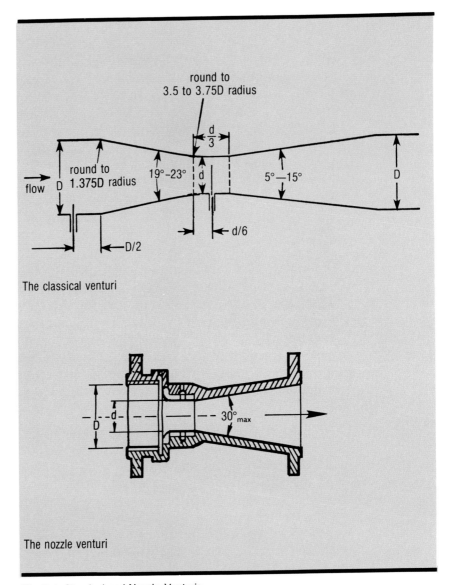

Fig. 5-4. Classical and Nozzle Venturis

Recommended dimensions in a ratio form are given in Fig. 5-5 for both the tube surfaces as well as the pressure taps.

Sizing of the classical venturi is accomplished in a manner similar to the orifice plate, except for one outstanding difference. The discharge coefficient for the venturi is known for a range of β and Reynolds number.

The discharge coefficient for a *rough-cast* converging cone venturi is:

$C = 0.984$

and is subject to a ±0.70% uncertainty when

  4 in (100mm) $\leq$ D $\leq$ 32 in (800 mm)
  $0.3 \leq \beta \leq 0.75$
  $2 \times 10^5 \leq R_D \leq 2 \times 10^6$

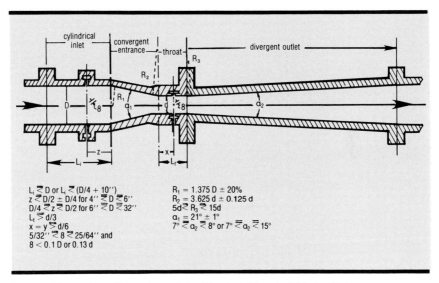

Fig. 5-5. Dimensional Proportions of the Classical (Herschel) Venturi Tube

For a *machined* entrance cone:

$C = 0.995$

subject to a ± % uncertainty when,

  2 in (50mm) $\leq$ D $\leq$ 10 in (250mm)
  $0.4 \leq \beta \leq 0.75$
  $2 \times 10^5 \leq R_D \leq 1 \times 10^6$

Rough-welded sheet-metal converging cone configurations have a discharge coefficient:

$C = 0.985$

with a ±1½% uncertainty when

  8 in (200mm) $\leq$ D $\leq$ 48 in (1200mm)
  $0.4 \leq \beta \leq 0.70$
  $2 \times 10^5 \leq R_D \leq 2 \times 10^6$

Consequently, no estimate for C, discharge coefficient of the classical venturi is required and the beta ratio β may be calculated directly as in Step 3 of the sizing exercise discussed earlier in this unit.

The proportions of the nozzled venturi are given in Fig. 5-6. Upstream pressure taps are corner taps and the throat taps, at least four, lead to an annular chamber or manifold.

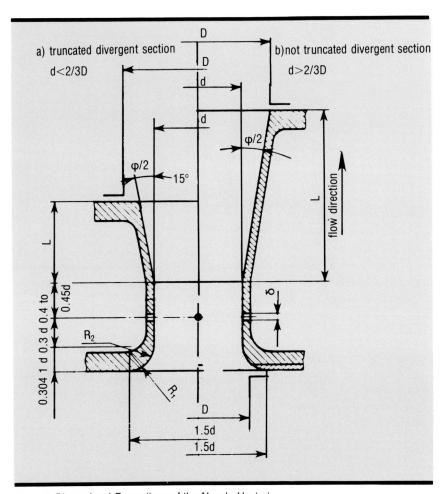

Fig. 5-6. Dimensional Proportions of the Nozzle Venturi

Venturi nozzles should be used only when:

$$2.5 \text{ in } (65\text{mm}) \leq D \leq 20 \text{ in } (500\text{mm})$$
$$d > 2 \text{ in } (50\text{mm})$$
$$0.316 \leq \beta \leq 0.775$$
$$1.5 \times 10^5 < R_D < 2 \times 10^6$$

Under these conditions the discharge coefficient, C, is given by the formula:

$$C = 0.9858 - 0.196\ \beta^{4.5} \tag{5-2}$$

5-3. Nozzles

For the nozzle venturi, a value of C = 0.98 should be chosen as a first approximation for the discharge coefficient in Step 3. After checking that the Reynolds number range for the desired application is compatible with the applicable range of the nozzle venturi (Step 5), then Eq. (5-2) is used to calculate the exact value of the discharge coefficient (Step 6).

Like the orifice and venturi, preferences for a specific type of flow nozzle differ between Europeans and Americans. In Europe, the International Standards Association (ISA) 1932 nozzle is preferred while in the U.S.A., preference goes to the ASME *long-radius* flow nozzle.

If you are getting the feeling that the conventional head-class flowmeter picture is streaked with controversy and opinion, you are absolutely correct. Advanced studies in head-class flowmetering should cover the nuances that lead the user to select one device over the other. Discussion on these points here is beyond the scope of this course.

As shown in Fig. 5-7, there are two recommended long-radius flow nozzles: the *high* β and *low* β nozzle. the term *long-radius* refers to the curvature of the inlet to the nozzle throat which, in reality, is the quadrant of an ellipse. Proportions of the nozzles are given in ratio form in Fig. 5-7. High β nozzles are recommended for use when the flow rate and desired differential pressure result in a β ratio falling between 0.45 and 0.80, whereas low β nozzles are recommended when the β ratio falls below 0.5.

In general, the advantages of the nozzle over the venturi are its shorter overall length and lower cost. The nozzle shares the same advantages of the venturi over the orifice plate, but the recovered pressure, although better than the orifice, is not as good as the venturi. The graph in Fig. 5-8 demonstrates this point.

Unit 5. Head-Producing Flowmeters I (Conventional)

Like the venturi, the discharge coefficient calculations for the nozzle are less complicated than for the orifice plate; however, it is more complicated than for the venturi. Discharge coefficients are the same for both types of long-radius nozzles and C is given by the following equation

$$C = 0.9965 - 0.00653 \, \beta^{1/2} \left[\frac{10^6}{R_D} \right]^{1/2} \qquad (5\text{-}3)$$

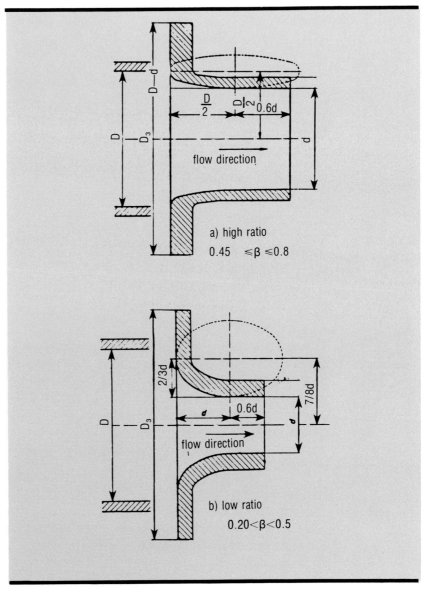

Fig. 5-7. Long-Radius Flow Nozzle

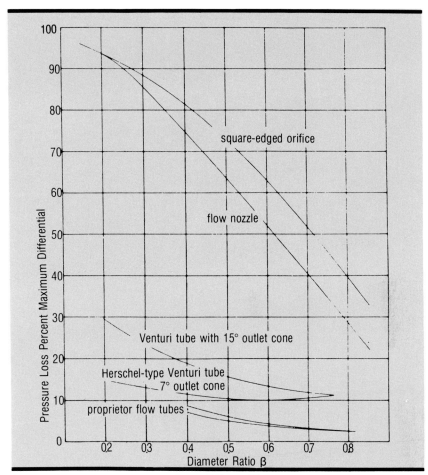

Fig. 5-8. Overall Pressure Loss for Several Primary Elements

when the following conditions are satisfied:

$$2 \text{ in } (50\text{mm}) \leq D \leq 25 \text{ in } (650\text{mm})$$
$$0.2 < \beta < 0.8$$
$$10^4 < R_D \leq 10^7$$

In Fig. 5-9, the cross-section of an ISA 1932 nozzle is shown at a plane passing through the center of the throat. Corner taps are used to measure the pressure drop across the nozzle.

The discharge coefficient C for the ISA 1932 nozzle is given by the following equation:

$$C = 0.9900 - 0.2262 \ \beta^{4.1} \qquad (5\text{-}4)$$
$$+ \left[0.000215 - 0.001125 \ \beta + 0.002490 \ \beta^{4.7} \right] \left[\frac{10^6}{R_D} \right]^{1.15}$$

when the following condition is satisfied:

$$2 \text{ in } (50\text{mm}) \leq D \leq 20 \text{ in } (500\text{mm})$$
$$0.3 < \beta < 0.8$$

and R_D is within the following limits:

for $0.30 \leq \beta \leq 0.44 \quad 7 \times 10^4 \leq R_D \leq 10^7$

for $0.44 \leq \beta \leq 0.80 \quad 2 \times 10^4 \leq R_D \leq 10^7$

The sizing steps employed for the orifice plate may be applied to the flow nozzle using the appropriate equations for the discharge coefficient, just as was done for the nozzle-venturi.

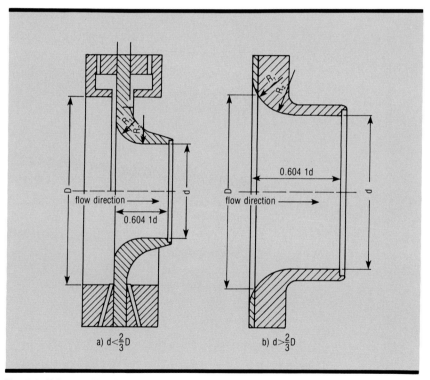

Fig. 5-9. ISA 1932 Nozzle

For flow nozzles, a value of C = 0.99 is chosen as a first approximation to the discharge coefficient in Step 3. After checking the Reynolds number range (Step 5), either Eq. (5-3) or Eq. (5-4) is used to calculate the exact value of the discharge coefficient for either the chosen long-radius nozzle or the ISA 1932 nozzle.

Exercises

5-1. Given the following process conditions:

- four-inch schedule 40 pipe (true D = 4.026 inches)
- Process fluid is water at 1.0 atmospheric pressure
- Process temperature: T = 90°F
- Maximum flow rate: $\dot{m}$ = 50 lb_m/sec
- Maximum head on D/P cell: Δh_{max} = 600 inches of water (at 90°F)
- Gravitational acceleration: g = 32.17 ft/sec²

Determine the following parameters:
 (a) The process fluid density, ρ, in lb_m/ft^3
 (b) The process fluid viscosity, μ, in lb_m/ft sec
 (c) The maximum fluid velocity, V_{max}, in ft/sec
 (d) the maximum differential pressure on D/P cell, ΔP_{max}, in lb_m/ft^2
 (e) Reynolds number based on the pipe diameter, R_D

Check your answers to Exercise 5-1 before proceeding to Exercise 5-2.

5-2. In this exercise you will size an orifice plate with **flange taps** for the flow situation described in Exercise 5-1.

 (a) Assuming that C = .6, find β (You can use Eq. 5-1 or Eq. 4-5a). Remember to round up to the nearest hundreth.
 (b) Calculate R_d. You can do this by dividing R_D by the β ratio.
 (c) Find the actual value of the discharge coefficient from Table 5-2.
 (d) Check this value using the Stolz equation.
 (e) Find the actual ΔP_{max}, using the Stolz equation value for the discharge coefficient. Express this result in inches of water (Δh_{max}). Is it lower than the original 600 inches?

5-3. In this exercise you will size a **rough-cast** venturi tube for the flow situation described in Exercise 5-1.
 (a) What is the flow coefficient for a **rough-cast** converging core venturi?
 (b) Find β.
 (c) Is the Reynolds number within an acceptable range for a **rough-cast** venturi? Is the β value within an acceptable range?

(d) Assume that a more sensitive pressure cell becomes available, one for which $\Delta h_{max} = 120$ inches of water. Using this Δh_{max}, what is the new β value? Is it within an acceptable range?

(e) Using the β ratio (rounded up ...), determine the actual Δh_{max} in inches of water. Is it lower than 120 inches?

5-4. In this exercise you will size a **long-radius** nozzle for the flow situation in 1, except that a 120-inch water D/P cell (as in 3) will be used.

(a) Assuming that C is approximately 0.99, find an approximate β.

(b) Using the approximate β from part (a), find a better approximation for C (using the relation for a **long-radius** nozzle).

(c) Part (b) gives a very good approximation of the flow coefficient. Use this new C value to determine β (round up ...).

(d) Are the Reynolds number and β ratio within acceptable range?

(e) Use the β ratio from part (c) to determine the actual flow coefficient (yes, again ...) and check that Δh_{max} is less than 120 in H_2O.

References

[1] Clemens, Herschel, "The Venturi Water Meter," *Transactions of the American Society of Civil Engineers*, 1887.

[2] Weymouth, Thomas R., "Measurement of Natural Gas," *Transactions of the American Society of Mechanical Engineers*, 1912, p. 1091.

[3] Froude, William, "Discharge of Elastic Fluids under Pressure," Proceedings, Institute of Civil Engineers (Great Britain), 1847, Vol. 6.

[4] Bean, Howard S., ed., "Fluid Meters, Their Theory and Application—Report of ASME Research Committee on Fluid Meters," American Society of Mechanical Engineers.

[5] "Measurement of Fluid Flow by Means of Orifice Plates, Nozzles, and Venturi Tubes Inserted in Circular Cross-Section Conduits Running Full," International Standards Organization 5167-1980 (E).

Unit 6: Head-Producing Flowmeters II (Special)

Unit 6

Head-Producing Flowmeters II (Special)

There have been some very innovative primary devices developed over the years, which are based on differential-pressure-producing principles, but have physical forms or behaviors more suited for use in certain application conditions not suited to the conventional head-producing device. These devices, all head-producing flowmeters, are classified as "special" because of their use in special instances or under special circumstances. In this unit, these devices are discussed from the applications point of view and the basic physics of operation of each device is explained.

Learning Objectives—When you have completed this unit, you should:

A. Understand the operating principles of several differently configured devices which produce a differential pressure as a measure of the flow rate.

B. Know when and why these devices are employed.

6-1. Pipe Elbow or Centrifugal Flowmeters

Centrifugal force is a physical phenomenon common to our everyday experiences in that, every time a turn is made in an automobile, the driver and passengers feel a force acting to move them in a direction along and outward on the turn radius. This acting force is called *centrifugal force*. Similarly, a fluid flowing in a curved channel exerts a higher force on the larger radius outer surface than on the smaller radius inner surface. A differential pressure is produced between opposite walls in a curved channel or pipe elbow, and the pressure differential or head follows the square-law behavior typical of head-class devices.

A sketch of a *pipe elbow* flowmeter is shown in Fig. 6-1. Advantages of the pipe elbow meter are the low pressure loss and the fact that most piping systems require an elbow, in any case, so that one may consider the pressure loss incurred as part of the piping system and the cost as necessary to the piping arrangement.

Disadvantages of the elbow flowmeter are the normally low differential pressures produced due to the dependence of the differential pressure on pipe diameter and elbow radius, which in turn leads to inaccuracies in the absolute measured flow rate. Consequently, the elbow flowmeter is more often used on flow control, where a repeatable measurement is of major concern.

In using an *elbow* flowmeter, at least 25 diameters of straight pipe upstream and 10 diameters downstream should be provided to ensure predictable performance. The elbow should be flanged and the diameter, D, should be the same as the approach pipe. An elbow smaller than the approach pipe *must* be calibrated and a short reducer section should be used to make the size change. Elbows with a diameter larger than the approach pipe, if used, must also be calibrated.

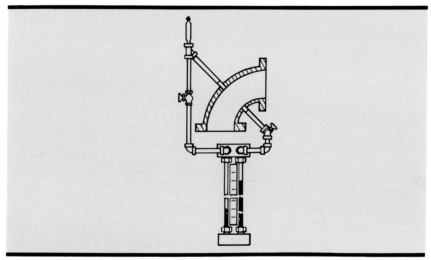

Fig. 6-1. Pipe Elbow Flowmeter

Usually, pressure taps are located at a 45° angle from each flange and are diametrically opposite at the middle and at the inside and outside of the bend. Flow may be in either direction when 45° taps are employed and most of the tests and experiences reported are for elbow flowmeters having 45° pressure taps. Some limited data indicate a differential pressure measured at 22 1/2° after the entrance to the elbow is more stable and less affected by upstream conditions because of a flow separation from the wall of the elbow beyond 22 1/2°, depending on the sharpness of the bend. Flow separation would create a highly noisy and unstable pressure signal because of the high level of turbulence present in the separated region.

Unit 6: Head-Producing Flowmeters II (Special)

A simplified version of the theoretical equation for the flow rate measurement arrived at by making several simplifying approximations is given below.

$$V_t = \sqrt{\frac{R\,\Delta P}{D\,\rho}} \qquad (6\text{-}1)$$

where:

- V_t = velocity, theoretical flow rate
- ΔP = differential pressure
- ρ = fluid density
- R = radius of elbow center line
- D = pipe diameter

Just as for conventional head-class meters, a discharge coefficient may be defined as:

$$C = \frac{\text{actual flow rate}}{\text{theoretical flow rate}}$$

Consequently,

$$C = \frac{V_a}{\sqrt{\dfrac{R\Delta P}{D\rho}}} \qquad (6\text{-}2)$$

Unlike the *conventional* head-class primary devices, the *pipe elbow* flowmeter has not been tested extensively; consequently, no standards of performance, fabrication, or geometric design have been sanctioned by national or international standards organizations. A review of published data does indicate, however, that for 90° elbows with pressure taps at 45°, a value of C is given by,

$$C = 1 - \sqrt{\frac{6.5}{R_D}} \qquad (6\text{-}3)$$

when $10^4 \leq R_D \leq 10^6$ and $\dfrac{R}{D} \geq 1.25$

Computed flow rates based on the value of C given by the above equation can be subject to an uncertainty of about ±4%. For flow-calibrated ebow meters, the accuracy should be comparable to that of the conventional type of head-class flowmeter.

6-2. Target Flowmeter

Meters of this type are also referred to as vane, gate, drag, or force flowmeters. A simple form of this type of meter is shown in Fig. 6-2. Fluid flowing through the device causes a plate to swing outward, away from the fluid jet. At no flow, the plate rests against the end section of the meter. In its simplest form, the device is used as a flow indicator only and no flow scale is provided. However, the meter may be provided with a scale for the determination of flow rate, where the scale might be the angle through which the *target* or vane swings.

Fig. 6-2. Vane-Type Flowmeter

A more sophisticated version of the *target*-type flowmeter is shown in Fig. 6-3. The device is comprised of a primary flow element, the *target* or disc mounted perpendicular to the flow direction, and a secondary element which measures the force on the target by means of a force-balance system. An annular orifice configuration is formed by the disc-shaped target in the circular pipe and the target is fixed to a rod which transfers the force to the force-balance mechanism.

As in the use of an orifice, the *target flowmeter* requires a laboratory calibration to establish a flow coefficient. The relationship between the force on the target, F, and the mass flow rate of the fluid is given by the following equation:

$$\dot{m} = C \left(\frac{D^2 - d^2}{d} \right) \sqrt{\frac{\pi}{2} g_c F \rho} \qquad (6\text{-}4)$$

where:

>D = pipe diameter
>d = target diameter
>g_c = gravitational constant (32.174 $\frac{lb_m}{lb_f} \frac{ft}{sec^2}$)
>F = force on target
>ρ = fluid density
>C = discharge coefficient

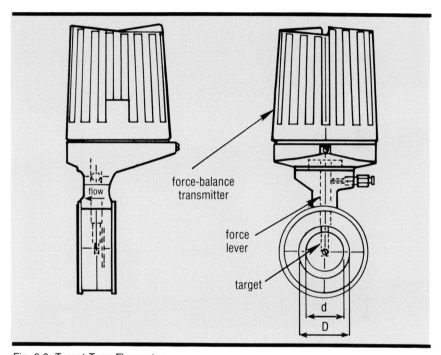

Fig. 6-3. Target-Type Flowmeter

Until recently, only limited coefficient data were available in the laminar flow regime. A study reported on in Ref. 1 was directed toward the determination of discharge coefficient data in the 100 to 2000 pipe Reynolds number range for one-half through four-inch-line size flowmeters. In addition, earlier data were included in the report that covers the transition and turbulent regions to pipe Reynolds numbers as high as 450,000.

An example of a typical discharge coefficient variation with pipe Reynolds number is shown in Fig. 6-4. Note the drastic reduction in the coefficient value at the low pipe Reynolds numbers. Consequently, care must be taken to generate experimental values for the discharge coefficient, especially if the meter is to be used at very low Reynolds numbers.

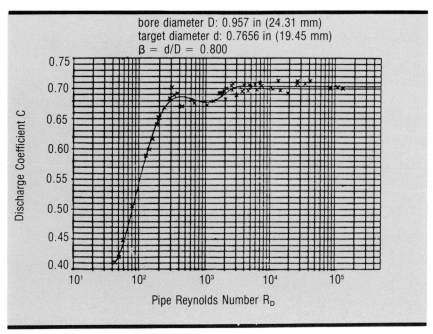

Fig. 6-4. Target Flowmeter Calibration Data

Target-type flowmeters originally were explored for application on dirty, sediment-laden and high-melting-point liquids that could not be measured by other head-class devices, like the orifice plate. It has been and still is known as the meter to specify when difficult applications arise and repeatability or precision is of main concern rather than absolute accuracy. However, with proper calibration and installation, the device provides accuracies comparable to other head-class flowmeters.

6-3. Pitot and Pitot-Static Probes

The *pitot tube* primary element is simply a tube having a short section of one end bent toward the flow direction, thereby allowing the flow to impact the tube opening, as shown in Fig. 6-5. A device of this type is also referred to as an *impact* tube.

In operation, the differential between the stagnation or impact pressure and the static pressure at the wall of the closed conduit is measured. The secondary element employed to measure the differential pressure may be simply a U-tube manometer or any one of the various forms of differential pressure measuring instruments and the differential pressure is proportional to the square of the velocity, making the device a *head-class* flowmeter.

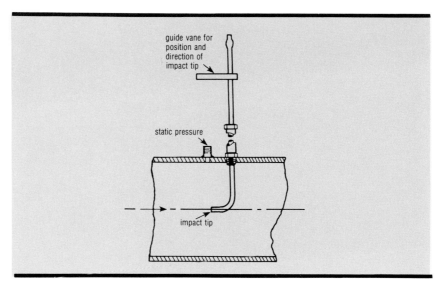

Fig. 6-5. A Basic Pitot or Impact Tube

A device of this type is normally employed in large conduits. Advantages of the device over other forms of flowmeter used for this application are its low cost and its slim form, which allows insertion into the conduit through a small opening. Disadvantages are its ability to measure the flow rate only at a point (the impact pressure point), its susceptibility to clogging in dirty fluids, and its sensitivity to the relative direction of the flow. To overcome these disadvantages, several innovative forms of the *pitot tube* concept have evolved over the years. For example, to overcome the disadvantage of a single-point measurement, a design of the type shown in Fig. 6-6 is available where several ports are placed along a tube in an attempt to measure an average impact pressure. In addition, a reversed impact tube is employed to alleviate the necessity for a separate static pressure port in the wall of the conduit. Although devices of this type are reported to be flow-rate averaging devices, they are affected by nonsymmetrical flow velocity profiles.

To overcome any sensitivity to flow direction, especially when the device is employed as an air speed indicator for aircraft, a shrouded probe arrangement shown in Fig. 6-7 is employed. In 1935, G. Kiel, Ref. 2, demonstrated that a venturi tube placed at an angle relative to the flow direction had internal streamlines parallel to the venturi axis up to angles of about $\mp 40°$. Consequently, a pitot-tube placed in the venturi should be insensitive to the flow direction. This probe is often referred to as a *Kiel probe* or *pitot-venturi*.

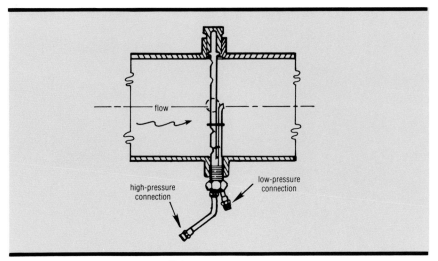

Fig. 6-6. Pressure-Averaging Pitot Tube

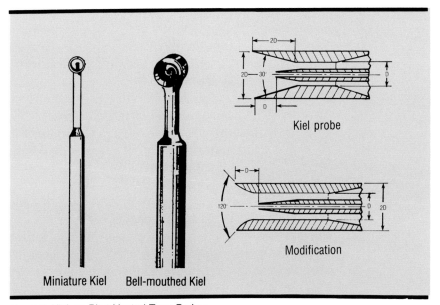

Fig. 6-7. Kiel- or Pitot-Venturi-Type Probes

Another advantage of this type probe is the high differential pressure obtained by measuring the differential between the impact pressure and the static pressure at the venturi throat, thereby allowing measurements of low velocity flows. A main disadvantage of these modified designs of *pitot tubes* is that they should be calibrated in the channel where they are to be used, or in one similar, whereas the conventional designs of *pitot* and *pitot-static* probes do not because their discharge coefficients are close to unity.

Two conventional and accepted designs for the *pitot-static* tube are shown in Fig. 6-8. The distinguishing feature of the *pitot-static* probe in contrast to the *pitot* probe is that the static-pressure measurement is an integral part of the probe design and is a true static-pressure measurement in close proximity to the impact-pressure measurement location. In design, two coaxial tubes are proportioned and assembled such that the inner tube senses the *impact* or *stagnation* pressure and the outer tube senses the *static* pressure through one or more rings of small holes in the wall of the tube and located well back from the tip.

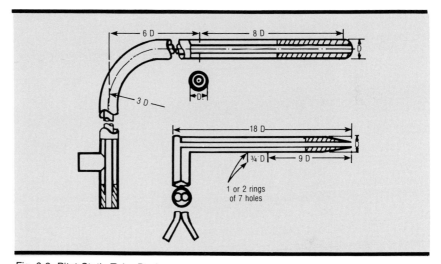

Fig. 6-8. Pitot-Static Tube Design

Other designs of pitot-static tubes are used for such special purposes as flow-direction measurement as well as flow-rate magnitude. A simple, rugged, two-dimensional direction probe configuration is comprised of a cylinder inserted normal to the plane of the angle measurement. As seen in Fig. 6-9, the three-hole direction probe is constructed with a stagnation pressure tap between the two direction taps. To provide for maximum sensitivity, the direction taps are located at 45° on either side of the stagnation-pressure tap. If the probe is rotated in the fluid stream until the pressure difference between the two direction taps is zero, the bisector of the angle between the holes gives the flow direction and the stagnation-pressure tap gives the true impact pressure. The differential between the impact pressure and the static pressure, as measured by one of the directional taps, provides for the true velocity measurement. A pitch angle of the flow stream (i.e., the angle δ in a plane parallel to the cylinder axis) does not affect the direction measurements, but does produce an error in the stagnation-and-static-pressure measurements. The error due to pitch in static pressure is less than 1% if the pitch angle is less than 5°. Sensitivity of the instrument has been determined as 5% of the stagnation-pressure per degree of yaw angle, ψ, which makes the device useful for measurements to within 0.1° at velocities over 100 feet per second in air.

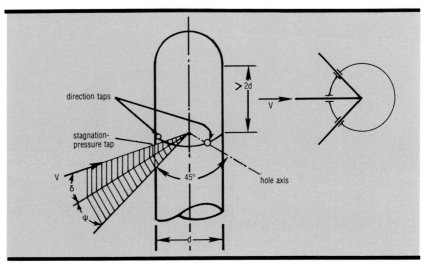

Fig. 6-9. Flow-Direction Measuring Pitot Probe

In computing the velocity using the pressure measurements taken with a *pitot* or *pitot-static* tube, the equation used is the same familiar form derived from Bernoulli's equation for an incompressible fluid.

$$V_1 = \sqrt{\frac{2\, P_s - P_1}{\rho_1}} \qquad (6\text{-}5)$$

where
V_1 = Velocity of the fluid stream at the location of the probe tip and in a direction coaxial with the tip

P_s = Impact, total or stagnation pressure measured at the probe tip.

P_1 = Static pressure—preferably corresponding to the static pressure in close proximity to the probe tip, i.e., the point of velocity measurement.

ρ_1 = Density of the fluid at the location of the static pressure P_1.

If the fluid is compressible, the equation takes the form:

where
$$V_1 = \sqrt{\frac{2\gamma}{\gamma-1} \frac{P_1}{\rho_1} \left[\left(\frac{P_s}{P_1}\right)^{\frac{\gamma-1}{\gamma}} - 1\right]} \quad (6\text{-}6)$$

γ = specific heat ratio of the gas as defined in Unit 4.

In 1922, M. Barker, Ref. 3, completed a study on the performance of pitot-static tubes at low Reynolds numbers. It was found that at low Reynolds numbers below 30, based on the radius of the stagnation-pressure port, the indicated stagnation pressure P_s was higher than expected. To compensate for the errors experienced at these low velocity conditions, a correction term was added to the Bernoulli equation in the form of an extra term.

$$\frac{P_s - P_1}{\frac{1}{2}\rho_1 V_1^2} = 1 + \frac{3}{R_e} \quad (6\text{-}7)$$

where
$$R_e = \frac{\rho V r}{\mu} \leq 30$$

r = pitot tube radius

The extra term represents an error which increases with decreasing velocity.

6-4. Area Flowmeters

Special forms of head meters, called *area flowmeters*, are constructed such that the area of flow restriction varies so as to hold the differential pressure constant. The volume flow rate is inferred by the change in area.

Common forms of the *area* meter are the *tapered tube* and *float*, the *cylinder* and *piston*, the *orifice* and *plug*, and the *rotameter*. A composite sketch of these flowmeter forms is provided in Fig. 6-10. All area-type flowmeters operate on the same basic physical principles, and the working equations for each type have the same form. In the operation of the *tube* and *float* devices, or *rotameter*, the float rises in the tube as the flow increases until the space between the tube and the float is large enough to decrease the fluid velocity to a value that results in a balance between the force of the flow acting on the float and the mass of the float. Because the devices are designed such that the differential pressure or head remains constant, the volume rate of flow is directly proportional to the space between the float and the walls of the tube. In addition, for a constant diameter float and a uniformly tapered tube, the area of the space is proportional to the height of the float in the tube.

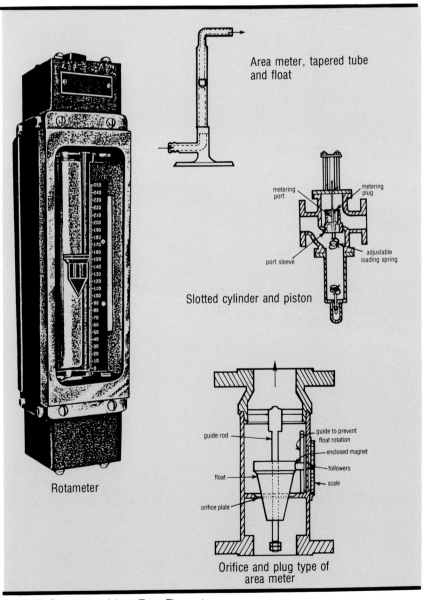

Fig. 6-10. Composite of Area-Type Flowmeters

The basic equation for area flowmeters has a general form for the expression of mass flow rate.

$$\dot{m} = (a^2 - 1) C \sqrt{2g_c AF\rho} \qquad (6\text{-}8)$$

where

> A = Area of tube corresponding to diameter D
> D = Tube diameter at elevation of float
> a = Ratio of the major to minor diameter of the variable annular metering areas, D/D_f
> D_f = Float diameter at reading surface
> C = Discharge coefficient
> g_c = gravitational constant (32.174 $\frac{lb_m}{lb_f} \frac{ft}{sec^2}$)
> F = Force (buoyed weight of float)
> ρ = Density of fluid

As for other head-class flowmeters, the actual flow rate, as measured by an *area* flowmeter, is not exactly equal to the theoretical flow rate and a discharge coefficient, C, is required to extract the actual flow rate from the theoretical prediction. Area meters are usually calibrated for use at a specific set of operating conditions and for a particular fluid. Thus, the discharge coefficient is determined over the desired range of operation. If this is not done, errors may occur due to the change in discharge coefficient resulting from changes in the fluid viscosity, for example. A thin disc float is almost independent of fluid viscosity over a wide range; however, the spherical or cylindrical shapes have significant viscosity sensitivity. Although advanced development in float designs has reduced the flowmeter's sensitivity to viscosity, a calibrated device is still preferable.

Tapered tube and *float* designs like the *rotameter* are generally available in sizes ranging from less than 1/16 in diameter to over 12 inches. Fluid velocities range from 1/2 to 10 feet per second. Minimum scale readability is nominally 10% of the maximum scale range.

Meters are supplied having the following features: remote indication and recording, operating temperatures up to 800°F, pressures up to 10,000 psi, and the ability to handle acids, alkalis, and even opaque fluids.

Some advantages of *area* flowmeters are the ease of installation, low maintenance, ease of reading, easy detection of faulty operation, low pressure loss, no flow straighteners are required, and no special approach or exit conditions are required. Disadvantages of the device are the requirement for installation in the vertical position, they are not suitable for dirty fluids or fluids with suspended solids, and the device is relatively fragile.

Cylinder and *piston* designs use a spring to provide a restoring force, as opposed to the weight, of a float; therefore, they are not normally considered to have a constant pressure loss. However, the operation of the flowmeter is similar to the *rotameter* in that the force of the fluid acting on the piston drives it vertically until the spring counteracts the force. Then the fluid flows out through holes or slots in the cylinder and the height of the piston is proportional to the flow rate.

Orifice- and *plug-*type devices are comprised of a tapered plug which rides vertically within the bore of an orifice. With increasing flow, the plug rises, thereby increasing the area of the orifice and allowing more flow to pass. When the mass of the plug is balanced by the force of the flow, the height of the plug becomes an indication of the flow rate.

6-5. Linear-Resistance Flowmeters

When measuring low flow rates in or near the laminar flow regime, a primary element is used which is designed to drive the flow further into the laminar regime, thereby taking advantage of the linear relationship between differential pressure and the rate of flow. This linear relationship is derived in great detail in Unit 4 and the final theoretical form of the laminar flow equation is given again below.

$$V_t = \frac{D^2 (P_1 - P_2)}{32 L \mu} \tag{6-9}$$

There are two common forms of primary elements for flowmeters designed to operate in the laminar flow regime: the *capillary tube* and the *porous plug,* shown in Fig. 6-11. The *capillary tube* is designed such that the small bore of the capillary *and* its long length make the flow more laminar in that both parameters present a resistance to the flow. Similarly, the *porous plug* containing such high-resistance materials as steel wool, glass wool, or layers of fine screening, is designed to accomplish the same task.

An important design factor in the capillary tube is the ratio of the length to the bore which should be greater than 150. Another design factor is the shape of the entrance and exit of the tube which should be smooth and well-tapered to eliminate turbulence. To increase flow capacity, a bundle of capillary passages is often used.

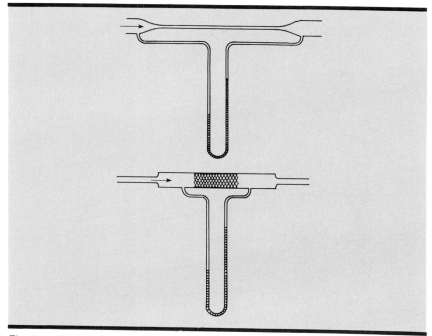

Fig. 6-11. Capillary Tube and Porous Plug

Porous plugs are made by packing a pipe or tube with a porous material and providing for a pressure-differential measurement across the plug. Important factors influencing the pressure drop are the compactness of the porous material and the length of the porous plug.

Referring to the equation describing the relation between the laminar flow rate and the pressure drop given above, we introduce the discharge coefficient using the same previously employed definition of this term, i.e., the ratio of actual to ideal (theoretical) flow rate.

$$C = \frac{V_a}{\dfrac{D^2 (P_1 - P_2)}{32\mu L}} \qquad (6\text{-}10)$$

where

D = Diameter of capillary tube
L = Length of tube between pressure measurement points
(P1 − P2) = Pressure drop
V_a = Actual velocity
μ = Fluid viscosity

This coefficient, obtained by calibration of a specific unit, counteracts any errors that are induced by inlet and exit losses or by construction anomalies. All laminar flowmeters should be calibrated; and remember, these devices all must operate in the laminar regime, which means the Reynolds number as referred to the metering passage should be less than 2000.

Exercises

6-1. *A simple piston-area flowmeter is shown in the sketch below. The area (A) of the flow stream is directly proportional to the height (h) of the opening.*

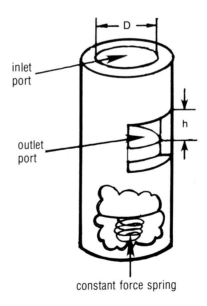

(a) For a meter such as this, a **meter factor** (K) is usually determined experimentally during a proof test. Which of the following relationships would you expect to observe?

1) $\dot{m} = K\sqrt{h}$
2) $\dot{m} = Kh$
3) $\dot{m} = Kh^2$

(b) If the spring exerts a force F on the piston, what will be the pressure drop in the meter?

(pressure = force/area)

6-2. Water (ρ = 62.4 lb$_m$/ft^3) flows through an elbow flowmeter at a rate of 2 cfs. The radius of curvature and the diameter of the pipe are six inches.
(a) Determine the differential pressure between the taps. (Assume C = 1.0)
(b) If the radius is doubled, what happens to the differential pressure? Calculate ΔP.

6-3. Water flows through a target flowmeter and produces a 0.40-pound force on the target. What is the mass flow rate? Use the calibration data and meter dimensions in Fig. 6-4, and ρ = 62.4 lb$_m$/ft^3 ν = 1.41 x 10^{-5} ft^2/sec

6-4. Water flows through a 2 mm dia. 30 cm. long capillary tube. The head loss measured across the capillary is 24 mm of mercury. What is the velocity of the water?

$$\left[C = 0.90, \mu_{H_2O} = 1.31 \times 10^{-3} \ \frac{N-s}{m^2}, \ \rho_{H_2O} = 1000 \ \frac{Kg}{m^3} \right.$$
$$\left. \rho_{Hg} = 13,600 \ Kg/m^3 \right]$$

6-5. A pitot-static probe is fastened to the hull of a boat and aimed forward. A mercury manometer shows a 1.5" differential head between the static and the stagnation taps from the probe. What is the speed of the boat relative to the water? (ρ_{Hg} = 13.55 ρ_{H_2O})

References

[1] Curran, D.E., "Laboratory Determination of Flow Coefficient Values for the Target-type Flowmeter at Low Reynolds Number Flow," *Flow, Its Measurement and Control in Science and Industry*, Vol. 2, 1981, ISA publication.
[2] Kiel, G., "Total Head Meter with Small Sensitivity to Yaw," NACA-TM775, August, 1935.
[3] Barker, M., On the Use of Very Small Pitot Tubes for Measuring Wind Velocity," *Proceedings of Royal Soc.*, Series A, Vol. 101, p. 435, 1922.

Unit 7:
Head-Producing Flowmeters III (Open Channel)

Unit 7

Head-Producing Flowmeters III (Open Channel)

The two basic types of head-class open-channel flow-measurement devices are the *weir* and the *flume*. Both types are "true" head-class flowmeters in that the flow rate is directly proportional to the head or height of liquid in the flowmeter. In this unit we shall discuss the basic principles of these "true" head-class flowmeters and explain the differences between *weirs* and *flumes*. Some interesting background reading on both devices may be found in Refs. 1 through 4.

Learning Objectives—When you have completed this unit you should:

A. Understand the differences in design and usage between weirs and flumes.

B. Know the several different configurations of weirs and flumes.

C. Be able to size an open-channel flowmeter for a simple application.

Open-channel head-class flowmeters are used only in conduits that contain a free fluid surface. Consequently, they are only used in the metering of liquids and are used almost exclusively in the metering of water flows. Two forms of open-channel flowmeters in use today are the *weir*, which is a dam over which the liquid flows, and the *flume*, which is a venturi-type section placed in an open channel.

Discussion of the basic terminology and operating principles of the weir will be presented first.

7-1. Weirs

Before a discussion on the various types of weir and the terminology associated with the different configurations, we will develop the basic flow equation.

Consider the flow of liquid over a dam as depicted in Fig. 1, where a stream filament of incremental thickness dy is defined.

The pressure acting on the streamline filament causing the subsequent fluid motion is expressed in terms of the fluid height y or *head*, as:

$$P = y\rho g \qquad (7\text{-}1)$$

This equation demonstrates the connection between pressure and fluid height or *head* and the origin of the term "true" head-class meter when referring to the *weir* or *flume*.

For both *weirs* and *flumes*, the pressure producing the flow is measured by the height of the free surface upstream of the dam face (usually called the *crest*). Use of the term *head* for the fluid height is universal and traditional. In contrast to the *conventional* class of head-producing flowmeters, where a pressure difference is measured, the measurement of *head* for a *weir* or *flume* represents a single pressure.

With the introduction of Bernoulli's equation and the expression of velocity in the form:

$$V = \sqrt{2\frac{P}{\rho}} \qquad (7\text{-}2)$$

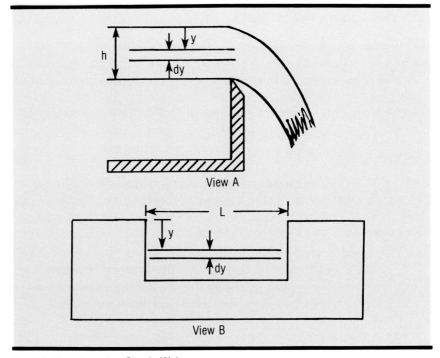

Fig. 7-1. Flow through a Simple Weir

one may then combine Eqs. (7-1) and (7-2) to arrive at an expression of velocity in terms of fluid height in the form:

$$V = \sqrt{2gy} \tag{7-3}$$

Now the incremental volume flow rate contained by the streamline filament may be expressed as:

$$dq = LVdy \tag{7-4}$$

where L is the width of the dam.

Substituting for V, the above equation is expressed in integral form as:

$$q_t = L\sqrt{2g} \int_o^h y^{1/2} dy \tag{7-5}$$

and upon integration for y over the height o to h we obtain:

$$q_t = \frac{2}{3} L \sqrt{2g}\, h^{3/2} \text{(theoretical flow rate)} \tag{7-6}$$

As for other flow meters we have discussed, the actual flow rate q_a is less than the theoretical and a coefficient of discharge is used, therefore:

$$q_a = \frac{2}{3} CL \sqrt{2g}\, h^{3/2} \text{(actual flow rate)} \tag{7-7}$$

where as before

$$C = q_a/q_t$$

The preceding development of Eq. (7-7) may be used to evolve the flow equations for the three most common designs shown in Fig. 7-2, namely, the *Cippoletti* or *trapezoidal notch*, the *rectangular notch*, and the *triangular* or *v-notch*. Differences in the flow equations for each device arise from the geometric differences in the notch shape. For example, in Eq. (7-5) the term L does not lie in the integral because the notch is rectángular and L is independent of y. However, for the V notch weir, L is dependent on y and the resulting integration results in a 5/2 power dependency on the fluid height as a function of flow rate.

Equations for each weir design will be presented later; however, before presenting the equations the terminology associated with *weirs* will be discussed.

With reference to Figs. 7-2 and 7-3, the basic terms referring to the notch shape are the *crest*, which is the bottom edge of the weir notch and sometimes referred to as the *sill*, the *crest width*, the distance along the *crest* between the sides and the *notch width*, the horizontal distance between opposite sides of the notch. In the design shown in Fig. 7-2, the *rectangular* and *Cippoletti weirs* have *crest width*, whereas the *v-notch* has zero *crest width* and variable *notch width*. Only the *rectangular weir* has constant *notch width*.

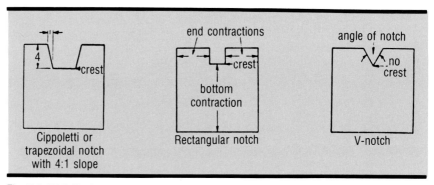

Fig. 7-2. Weir Designs

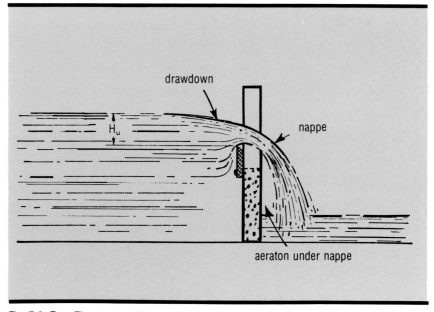

Fig. 7-3. Free Flow over a Weir

Referring to Fig. 7-1, the *head* of a *weir* is the vertical distance measured from the *crest* level to the liquid surface. The *head* measurement should be made at a distance of four or more times the maximum expected *head* value and at a location upstream of and measured from the weir plate. The sheet of liquid passing through the notch and falling over the weir crest is called the *nappe*. The liquid surface downstream of the weir plate should be far enough below the crest so that air has free access beneath the nappe. For this condition, the flow is said to be *free*. If this condition does not exist and no air is contained beneath the nappe, the flow is said to be *submerged*. The amount of submergence is measured from the crest level, just as the head measurement is made.

Drawdown is the curvature of the liquid surface upstream of the weir plate. The head measurement should be no closer to the weir plate than the upstream zero curvature point on the *drawdown*.

Contraction is the narrowing of the stream of liquid passing through the notch. The spacing between the walls of the upstream channel and the edge of the notch control the amount of *contraction*. For example, the amount of contraction is defined by the horizontal distance from the end of the crest to the side walls, called *end contractions* and the vertical distance from the crest to the floor of the weir box or channel bed, called the *bottom contraction*. A weir having both *end* and *bottom contractions* is said to have *complete contraction*. For a completely contracted weir, the stilling basin ahead of the weir plate should be large enough to pond the liquid so that it approaches the weir plate at low velocity. This stilling basin is called the *weir pond*.

For a *rectangular weir* in which the width of the approach channel is equal to the *crest width*, i.e., there are no *end contractions*, the weir is called a *suppressed weir*.

A weir should be sized to give a head measurement in excess of 0.1 feet at all flow rates of interest. If large flow rates are to be measured, the choice will usually be confined to either the *rectangular* or *trapezoidal weir*. For small flow rates, the choice is usually limited to the *v-notch weir* because of the 0.1-foot head measurement limitation. To ensure that a chosen measuring device delivers a usable range of heads at maximum and minimum expected flow, refer to Fig. 7-4 as a sizing guide.

Unit 7: Head-Producing Flowmeters III (Open Channel)

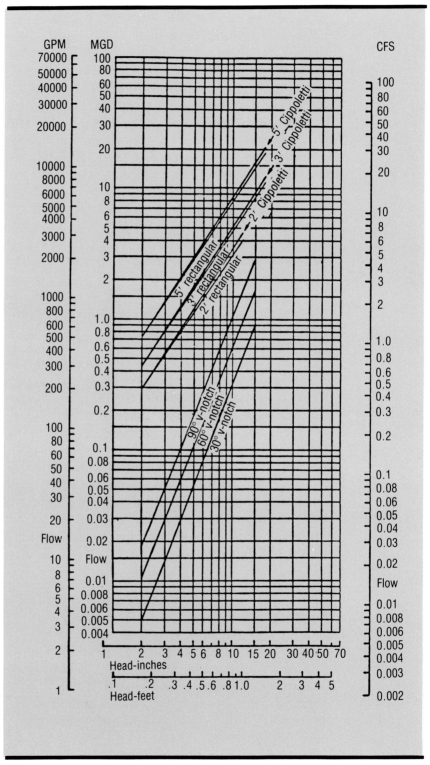

Fig. 7-4. Size Guide for Weirs

A simple *weir box* may be constructed using the composite dimension of Table 7-1 and the sketch in Fig. 7-5. For a rectangular weir with complete contractions, the width of the box should be three times the width of the notch. For maximum accuracy, the end contraction should also be three times the maximum head and the bottom contraction twice the head; or the end contraction may be twice the head and the bottom contraction three times the head. A *crest width* of less than 1/2 foot should not be used; therefore, when flows are small enough to require a small weir the *triangular* or *v-notch* are used.

Upstream edges of the weir plate should be sharp and straight. The usual practice is to level off the downstream edge of the weir at a slope of 45° to about 1/32-inch edge thickness. A triangular notch should be straight and sharp at the apex. The area of the approach channel or weir pool should be at least nine times the area of the notch at maximum head.

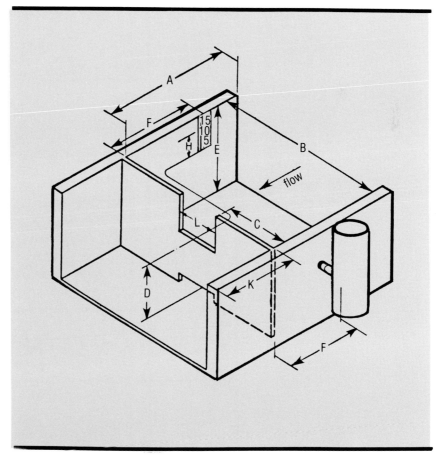

Fig. 7-5. Dimensions for the Weir Box

Weir-Box Dimensions for Rectangular, Cippoletti, and 90° Triangular-Notch Weirs

Rectangular and Cippoletti Weirs

Approximate limits of discharge	H Maximum head	L Length of weir crest	A Length of box above weir crest	K Length of box below weir crest	B Total width of box	E* Total depth of box	C Distance from end of crest to side of box	D Distance from crest to bottom of box	F Gauge distance
Second-Feet	Feet	Feet	Feet	Feet	Feet	Feet	Feet	Feet	Feet
1/10 to 3	1	1	6	2	4	3	1 1/2	1 1/2	4
1/5 to 6	1 1/4	1 1/2	7	3	5	3 1/4	1 3/4	1 1/2	4 1/2
1/4 to 8	1 1/4	2	8	4	6	3 1/2	2	1 3/4	5
1/3 to 17	1 1/2	3	9	5	7	4	2	2	5 1/2
1/2 to 23	1 1/2	4	10	6	9	4	2 1/2	2	6
3/4 to 35	1 1/2	6	12	6	11 1/2	4 1/2	2 3/4	1 1/2	6
1 to 50	1 1/2	8	16	8	14	4 3/4	3	2 3/4	8
1 to 60	1 1/2	10	20	8	17	5	3 1/2	3	8
90° Triangular-Notch Weir									
1/10 to 2 1/2	1	—	6	2	5	3	—	1 1/2	4
1/10 to 4 1/3	1 1/4	—	6 1/2	3	6 1/2	3 1/4	—	1 1/2	5

*This distance for about six inches freeboard above highest water level in weir box.

Table 7-1. Weir-Box Dimensions

Other general precautions are that when the flow rates are low, the velocity must be sufficient to overcome the surface tension so that the nappe is free. If this condition does not exist, the readings are unreliable. Also, submerged flows over weirs are not measured accurately because the flow rate depends on the downstream head as well as the upstream head.

Generally accepted practice in the use of weirs is to limit the maximum head to not more than 1/3 the crest width; however, laboratory experiments show the accuracy of the measurement is not impaired by exceeding this limit for a crest width between one and four feet. Usually heads of less than 0.1 foot or more than one foot should be avoided. The discharge coefficient curve begins to deviate from the predicted values for heads less than 0.2 feet, and when the head reaches one foot or more there is a tendency to form vertical boils or eddies in the approach section which tends to reduce the rate of flow over the crest. The velocity of the liquid in the weir pond upstream of the weir plate should not exceed 1/3 foot per second.

The accuracy of flow measurements made using weirs depends on many factors, such as the sharpness of the weir edge, the exactness of the weir dimensions and the accuracy of the head measurement. In addition, if the flow equation requires a discharge coefficient, there is the question of how closely the weir system corresponds to the system from which the coefficient data were obtained. Consequently, weirs are not considered high-accuracy devices and most reports on tests of weirs claim an accuracy of between $\pm 1/2$ to $\pm 4\%$ at a 75% confidence level. On a 95% confidence level, the uncertainty could be as high as ± 1 to $\pm 9\%$.

The equations for calculating the flow through weirs are listed below. These equations are generally accepted and are believed to give reasonably accurate measurements for the flow rate through the particular weir and for the conditions for which they are recommended.

Rectangular Weirs

a. Complete contractions and negligible velocity of approach

$$q_a = 3.33(L - 0.2H_u)H_u^{3/2}$$

b. End contractions suppressed

$$q_a = \left[3.228L + 0.4347 \frac{H_u}{H_b - H_u} \right] (H_u + 0.0036)^{3/2}$$

Cippoletti or Trapezoidal-Notch Weirs

Full end and bottom contractions:

$$q_a = 3.08L^{1.022}H_u^{(1.46 + 0.003L)} + 0.6H_u^{2.6}$$

or:

$$q_a = 3.367LH_u^{3/2}$$

Triangular or V-Notched Weirs

$$q_a = (0.025 + 2.462U)H_u^{(2.5 - 0.0195/U^{0.75})}$$

$$q_a = 4.28UCH_u^{5/2}$$

for 60° weir $q_a = 1.432H_u^{5/2}$
for 90° weir $q_a = 2.481H_u^{5/2}$ $\Big\}$ for C = 0.58

Symbols

q_a = actual rate of flow, cu ft/sec

L = length of crest, ft

H_u = upper head measured from crest, ft

H_b = lower head measured from bottom of weir box, ft

H = velocity head in weir pond
$\frac{(velocity)^2}{2g}$, where the velocity is expressed in ft/sec
and g is the acceleration due to gravity in ft/sec²

N_c = number of end contractions

U = slope of sides of notch from vertical or tan θ, when θ is one-half the included angle of the notch

C = experimentally determined discharge coefficient

7-2. Flumes

Generally speaking, the *flume* is an adaptation of the venturi concept of flow constriction applied to open-channel flow measurement. A flume is configured by contracting the side walls and by raising a section of the floor to form a low, broad-crested weir contour. In sharp contrast to the weir, flumes may operate with considerable submergence and low overall head. Also, being essentially a flow-through device as opposed to a dam type like the weir, the presence of silt, tree branches, etc., is not detrimental to performance.

Three types of flumes most commonly used are the *Parshall flume*, the *Palmer-Bowlus flume*, and the *parabolic discharge flume*. A major difference between the *Parshall flume* and the other two is that only the Parshall flume is used in rectangular, open-channel flow. Both the Palmer-Bowlus flume and the parabolic discharge flume are used in partially filled circular pipes; however, the parabolic flume must operate without submergence.

Let's consider first the application of the Parshall flume to rectangular, open-channel flow.

Parshall Flume

The Parshall flume was developed at the Colorado Agricultural Station in cooperation with the Division of Irrigation, U.S. Department of Agriculture.

As seen in Fig. 7-6, there are three basic parts to the flume configuration, a *converging section*, a *throat section*, and a *diverging section*. The *converging section* is the entrance to the flume throat. Both side walls are equally inclined inward and the floor is level. At the *throat section*, both walls are parallel to each other and vertical. The floor slopes downward. After the throat, the *diverging section* has vertical sidewalls inclined outward and the floor is inclined upward. For the Parshall flume, the *crest* is defined as the floor line forming the junction between the converging section and the throat. The size of the flume is determined by the length of the crest, L, and is identical to the normal distance between the walls of the throat.

In contrast to the weir, a head measurement may be used at both an upstream, *upper head*, H_u and downstream *lower head*, H_b, position when the *submergence* $(H_b/H_u) \times 100$ is greater than 70%. The measured rate of flow is in error if only the upper head measurement is used. The *upper head*, H_u, is measured at a point 2/3 the length of the converging section back from the end of the crest for flumes up to 10 feet in size. From 10 to 50 feet in size, H_u is observed at a position equal to $1/3 (L + 8)$ as measured along the side wall of the flume. The *lower head*, H_b, is observed at a point near the lower end of the throat section.

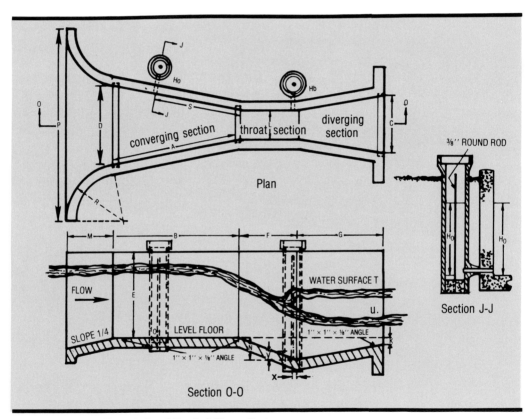

Fig. 7-6. Small Parshall Flume Design

Free flow through a Parshall flume is defined as the condition where the ratio H_b to H_u does not exceed 0.6 for flume sizes less than one foot or 0.7 for flume sizes one to eight feet. The lower head H_b is neglected for free-flow conditions. Consequently, the degree of submergence may approach 70% before the free-flow rate of discharge is affected. However, through the use of the two heads H_u and H_b, rate of flow may be computed with reasonable accuracy for up to about 95% of submergence.

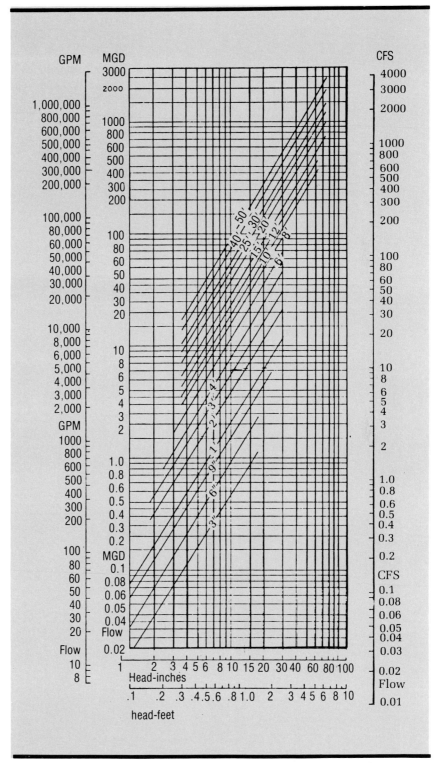

Fig. 7-7. Size Guide for Parshall Flumes

In the design of a flume, the volume amount of liquid per unit time as well as the elevation of the liquid surface in the channel should be known or defined quantities. With the use of Fig. 7-7 and the known volumetric flow rate, the size of the flume best suited to the particular situation is determined. It is important to note that the Parshall flume is categorized into small flumes and large flumes, depicted in Figs. 7-6 and 7-8. The pertinent dimensions for both are listed in Table 7-2.

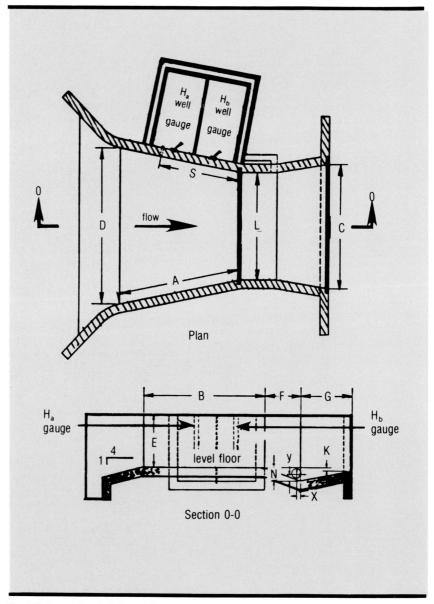

Fig. 7-8. Large Parshall Flume Design

Dimensions and Capacities of Small Parshall Measuring Flumes

L	A	S	B	C	D	E	F	G	K	N	R	M	P	X	Y	Free-flow Capacity Min.	Free-flow Capacity Max.
Ft. In.	Ft. In.	Ft. In.	Ft. In.	Ft. In.	Ft. In.	Ft. In.	Ft. In.	Ft. In.	In.	In.	Ft. In.	Ft. In.	Ft. In.	In.	In.	c.f.s.	c.f.s.
0—3	1—6⅜	1—¼	1—6	0—7	0—10 3/16	2—0	0—6	1—0	1	2¼	1—4	1—0	2—6¼	1	1½	.03	1.9
0—6	2—0 7/16	1—4 5/16	2—0	1—3⅝	1—3⅝	2—0	1—0	2—0	3	4½	1—4	1—0	2—11½	2	3	.05	3.9
0—9	2—10⅝	1—11⅛	2—10	1—3	1—10⅝	2—6	1—0	1—6	3	4½	1—4	1—0	3—6½	2	3	.09	8.9
1—0	4—6	3—0	4—4⅞	2—0	2—9¼	3—0	2—0	3—0	3	9	1—8	1—3	4—10¾	2	3	.11	16.1
1—6	4—9	3—2	4—7⅞	2—6	3—4⅜	3—0	2—0	3—0	3	9	1—8	1—3	5—6	2	3	.15	24.6
2—0	5—0	3—4	4—10⅞	3—0	3—11½	3—0	2—0	3—0	3	9	1—8	1—3	6—1	2	3	.42	33.1
3—0	5—6	3—8	5—4¾	4—0	5—1⅞	3—0	2—0	3—0	3	9	1—8	1—3	7—3½	2	3	.61	50.4
4—0	6—0	4—0	5—10⅝	5—0	6—4¼	3—0	2—0	3—0	3	9	2—0	1—6	8—10¾	2	3	1.3	67.9
5—0	6—6	4—4	6—4½	6—0	7—6⅝	3—0	2—0	3—0	3	9	2—0	1—6	10—1¼	2	3	1.6	85.6
6—0	7—0	4—8	6—10⅜	7—10	8—9	3—0	2—0	3—0	3	9	2—0	1—6	11—3½	2	3	2.6	103.5
7—0	7—6	5—0	7—4¼	8—0	9—11⅜	3—0	2—0	3—0	3	9	2—0	1—6	12—6	2	3	3.0	121.4
8—0	8—0	5—4	7—10⅛	9—0	11—1¾	3—0	2—0	3—0	3	9	2—0	1—6	13—8¼	2	3	3.5	139.5

Dimensions of Large Parshall Measuring Flumes

L	A	S	B	C	D	E	F	G	K	N	X	Y	Free-flow Capacity Min.	Free-flow Capacity Max.
Ft. In.	Ft. In.	Ft. In.	Ft. In.	Ft. In.	Ft. In.	Ft. In.	Ft.	Ft.	In.	Ft. In.	In.	In.	c.f.s.	c.f.s.
10—0	14—3⅜	6—0	14—0	14—8	15—7¼	4—0	3	6	6	1—1½	12	9	6	200
12—0	16—3¾	6—8	16—0	18—4	18—4¾	5—0	3	8	6	1—1½	12	9	8	350
15—0	25—6	7—8	25—0	24—0	25—0	6—0	4	10	9	1—6	12	9	8	600
20—0	25—6	9—4	25—0	29—4	30—0	7—0	6	12	12	2—3	12	9	10	1,000
25—0	25—6	11—0	25—0	34—8	35—0	7—0	6	13	12	2—3	12	9	15	1,200
30—0	26—6	12—8	26—0	45—4	40—4¾	7—0	6	14	12	2—3	12	9	15	1,500
40—0	27—6	16—0	27—0	56—8	50—9½	7—0	6	16	12	2—3	12	9	20	2,000
50—0	27—6	19—4	27—0		60—9½	7—0	6	20	12	2—3	12	9	25	3,000

Table 7-2. Parshall Flume Dimensions

The first step in the design procedure is to determine the size of flume. Therefore, if the maximum volume flow rate is 10 cfs and the water depth at this flow rate is one foot, a flume size of two feet is found (using Fig. 7-7) to be sufficient to handle the flow and deliver a reasonable head. After the size is determined, by reentering Fig. 7-7, it is found that for a 10-cfs flow rate and a two-foot flume, the expected upper head H_u is approximately 1.2 feet. To be conservative, it is usual practice to design for a 60% submergence because the submergence value may increase with time due to sand deposits filling in the downstream channel. Consequently, the lower head H_b is (1.2 × 0.6) or approximately 0.7 feet and the overall loss of head ($H_u - H_b$) is 0.5 feet. Because the lower head H_b may be neglected (considered unchanged) for flumes operating in the *free flow* condition, it is not a significant design parameter other than for the calculation of the *submergence*. Therefore, to reduce the loss of head, the size of the flume could be increased; however, if the flume size is increased, the degree of submergence becomes higher and the upstream head is lower. Consequently, the trade-off in design to optimize efficiency is between the upstream measuring head and overall head loss.

To conclude the discussion on the Parshall flume, the flow equations for a range of flume sizes are given below:

L = 3 in	$Q = 0.992\, H_u^{1.547}$
L = 6 in	$Q = 2.06\, H_u^{1.58}$
L = 9 in	$Q = 3.07\, H_u^{1.53}$
L = 1 ft to 8 ft	$Q = 4L\, H_u^{1.522}$
L = 8 ft and larger	$Q = (3.6875\, L + 2.5)H_u^{1.6}$

The symbols and units used here are the same as for the equations for a weir presented earlier in this unit.

Palmer-Bowlus Flume

The underlying principle of the Palmer-Bowlus flume is the establishment of a sufficient constriction to ensure that the kinetic energy, $1/2\, \rho V^2$, at the throat is greater than the kinetic energy in the free flowing conduit, for all values of expected flow rate. Consequently, the Palmer-Bowlus flume is applied to unusual flow situations and particularly to flows in circular conduits such as storm drains or sanitary sewers.

As shown in Fig. 7-9, both sidewall and bottom contraction are used to increase the kinetic energy at the throat and the throat length is equal to or greater than the diameter of the conduit.

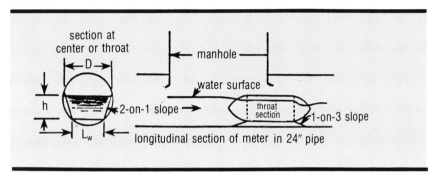

Fig. 7-9. Palmer-Bowlus Flume

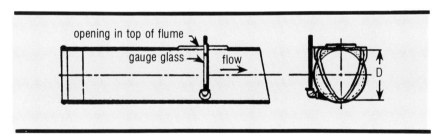

Fig. 7-10. Parabolic Discharge Flume

The equation describing the volumetric flow rate is similar in form to the weir or flume and is given by:

$$Q = Ch^{3/2} \left[\frac{(2L_w + h)^3 g}{8(L_w + h)} \right]^{1/2}$$

where the discharge coefficient C is usually determined by model calibration.

Parabolic Discharge Flume

The parabolic discharge flume is used on the outflow end of a pipe as shown in Fig. 7-10. A major difference between this design and the two flume designs discussed previously is that the discharge from this flume must fall free from the end of the conduit.

Simple in construction, the flume is formed by constricting the sides to form a parabola whose apex is in line with the bottom of the pipe. The flume size is equivalent to the diameter of the conduit on which it is used and the whole structure is mounted on the discharge end of the pipe.

As for the Palmer-Bowlus flume, the relation between fluid depth and flow rate is established by calibration.

Numerical values of the measurement accuracy of flumes are not readily available, primarily due to the lack of a primary calibration standard for use at such high volumetric flow rates normally encountered in irrigation or sewerage handling. Probably, flow measurements with flumes are somewhat less accurate than with weirs; however, there is no telling how less accurate. Consequently, the best course of action is to recommend a calibration, even of a scale model, when a specified accuracy is required.

Exercises

7-1. (a) Using Fig. 7-4, select a weir that creates at least a four-inch head for a flow of 0.05 cfs.
(b) Within the accepted limitations, what is the maximum flow rate suggested for a five-foot rectangular weir?

7-2. A completely contracted rectangular weir has a crest width of two feet. The head on the weir is six inches. What is the discharge?

7-3. Derive an expression for the theoretical relationship between flow and head for a 60° triangular weir.

7-4. An eight-foot Parshall flume operates with an upper head of 1.5 feet and 66.7% submergence.
(a) Calculate the lower head.
(b) Determine the volume flow rate through the flume.
(c) Calculate the head loss.

7-5. (a) What size Parshall flume will operate with 40% submergence, an upper head of 4.25 feet, and a volume flow rate of 1000 cfs?
(b) What is the head loss?

References

[1]Kindsvater, C.F. and Carter, R.W., "Discharge Characteristics of Rectangular Thin-Plate Weirs," *Transactions of the ASCE*, Vol. 124, 1959, p. 772.
[2]Smith, E.S., "The V-Notch Weir for Hot Water," *Transactions of the ASME*, Vol. 56, 1934, p. 787; and Vol. 57, 1935, p. 249.
[3]Parshall, R.L., "The Parshall Measuring Flume," Colorado Agricultural Experimental Station Bulletin 423, March 1936.
[4]Palmer, H.K. and Bowlus, F.D., "Adaptation of Venturi Flumes to Flow Measurement in Conduits," *Transactions of the ASCE*, Vol. 101, 1936, p.1195.

Unit 8:
Pulse Producing Flowmeters

Unit 8

Pulse Producing Flowmeters

The second class of flowmeters using the extractive energy approach to flow measurement is the *pulse-producing class*. Three types of flowmeters are included in this class, namely, the *positive-displacement type*, the *current type*, and the *fluid-dynamic type*.

Learning Objectives—When you have completed this unit, you should:

A. Understand the difference between the positive-displacement, current, and fluid-dynamic types of pulse-class flowmeters.

B. Be able to choose the pulse-class type best suited to the application.

8-1. Positive-Displacement-Type Flowmeters

The principle of displacement came to light in the third century B.C. when Archimedes observed that "any shape object will displace its volume of fluid when submerged." Conversely, a discrete volume of fluid will displace or move a solid body.

One of the distinctive features of a positive-displacement flowmeter is the passage of fluid through the primary device in discrete or isolated quantities. The number of discrete quantities is counted and flow indication is usually expressed in volume terms. To preserve completely (or almost completely) isolated quantities, two types of seal arrangements are normally employed: the *positive seal* or the *capillary seal*. A *positive seal* may be a flexible seal material (such as water, for example) or a pack seal or mechanical seal. In any case, the positive seal is a tight closure preventing any momentary fluid leakage into or out of the isolation chamber. The *capillary seal* provides sealing through the surface tension of a film of fluid existing between two surfaces of an isolation chamber that are not in actual physical contact.

A term commonly used to express leakage in positive-displacement flowmeters is *slip*. The flow through clearance or *slip* is defined as the difference between the measured fluid volume per cycle passing through the flowmeter and the calculated displacement per cycle determined from isolation chamber dimensions.

One of the earliest forms of positive-displacement flowmeters emerged in the gas industry. Samuel Clegg of England invented a revolving *liquid-sealed drum-type* meter in 1816, which was the first practical gas flowmeter. In the design, shown in Fig. 8-1, a cylindrical chamber is divided into rotating compartments formed by vanes. The chamber is filled more than half full of water, which acts as the flexible, positive-seal material. As gas enters through the center shaft into one compartment after another, the forces produce a rotation that allows the gas to exhaust out of the top as it is displaced by the water. Indicators, such as simple electric contacts or magnatic sensors, transmit the number of pulses per revolution of the drum.

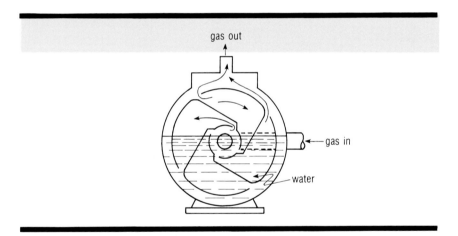

Fig. 8-1. Liquid-Sealed Drum Gas Meter

At one time, meters of this type were used by almost every gas manufacturing plant to measure the volume of gas made and delivered. To accomplish this task, some of the meters were constructed to be as large as 16 feet in diameter and proportionally as long.

In 1843, William Richards made a dry gas meter or *bellows meter* having two diaphragms, two slide valves, and a dial indicator. Later, Thomas Glover introduced the "Glover two-diaphragm slide valve meter," which remains practically unchanged even to the present date. The primary elements of both meters are the flexible diaphragms of the isolation chambers, valves for controlling the gas flow in filling and emptying the chambers, mechanical linkage to keep the diaphragms and valves synchronized, and a sensor to count the number of cycles.

In Fig. 8-2, a schematic diagram is presented which contains the stages of filling and emptying of four isolation chambers of a two-diaphragm meter. To maintain continous flow and insure sufficient power to operate a mechanical register, three or more chambers usually are used in the *bellows* form of positive-displacement flowmeter.

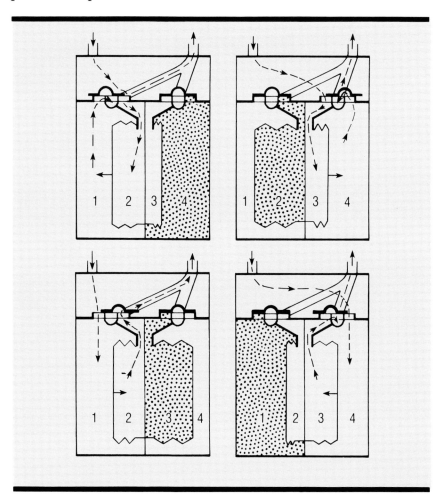

Fig. 8-2. Two-Diaphragm Dry Gas Meter

Another form of positive-displacement flowmeter is the *reciprocating-piston* mechanism, shown in Fig. 8-3. As fluid enters the area surrounding the crankshaft, the position of the pistons is such that inlet port I (3) is open to allow fluid to act on the face of piston (3) and simultaneously drive piston (1). Inlet port I (1) is open to the exhaust port E (1), therefore piston (1) is free to displace the fluid contained in the isolation chamber through the exhaust port E (1).

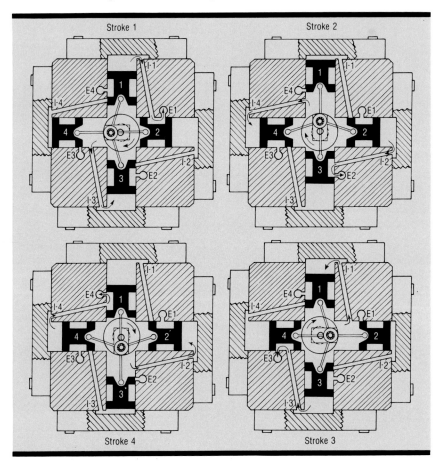

Fig. 8-3. Reciprocating Piston

At the end of the first stroke, inlet port I (4) is opened to drive piston (4) connected to piston (2). Fluid in the isolation chamber of piston (2) is then discharged through inlet port I (2) to the exhaust port E (2).

At the end of the second stroke, inlet port I (1) is open driving piston (1) and (3) with exhaust through E (3) during the third stroke. During the fourth stroke, fluid is displaced through port E (4) completing one complete cycle of rotation.

In reciprocating piston meters, mechanical sealing rings usually are used to eliminate slippage past the piston; however, capillary sealing is sometimes achieved with long pistons having a greater surface area. If capillary film sealing is employed, the flowmeter should be subjected to relatively low pressure drop across the seal by reducing to very low levels the load from the bearings and mechanical leakage.

Another form of piston meter, the *rotary piston* meter, is shown in Fig. 8-4. In this form, the outer surface of the circular piston (P) is always in contact with the inner surface of the cylinder (C) and the inner surface of the piston (P) is in contact with the outer surface of the cylinder (D). Sealing of the isolation chambers is achieved through sliding contact between the cylinder base and the end of the piston as well as a rolling and sliding contact between the cylinder and piston walls.

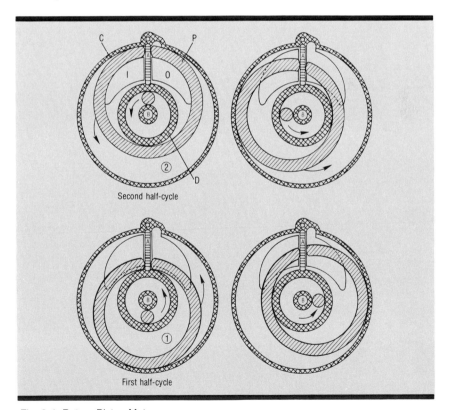

Fig. 8-4. Rotary Piston Meter

The kidney-shaped areas (I) and (O) are the inlet and outlet ports, respectively. Between the cylinders (C) and (D) there is a partition that prevents the rotation of the piston and provides a seal between the inlet (I) and the outlet (O) ports. The motion of the piston is shown in Fig. 8-4. Note that the piston also functions as a valve. In one half-cycle, fluid fills chamber (1) and in the second half-cycle chamber (1) is exhausted through the output port (O) and chamber (2) is filled for subsequent exhaust in the next half-cycle.

The *nutating disk* flowmeter is a novel and cleverly designed mechanical device. As shown in Fig. 8-5, the meter contains a circular disk attached to a spherical center. The disk is restrained from rotation by a vertical partition attached to the disk and installed in a slot; however, the disk is allowed to nutate (a circular precession or wobble motion).

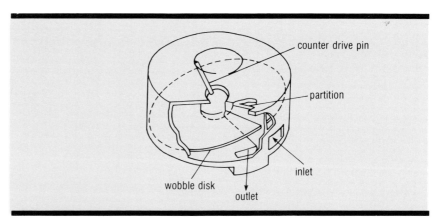

Fig. 8-5. Nutating Disk

In operation, fluid enters the isolation chamber alternately above and below the disk and proceeds around the conical chamber to the outlet port. The inlet and outlet ports are in the side wall of the case, on opposite sides of the vertical partition, which acts as a seal. Fluid motion around the isolation chamber, first above and then below the disk, produces the nutating motion of the disk. As the disk nutates, a shaft mounted through the center of the sphere moves in a circular path and actuates the meter register.

Several forms of the *rotary-vane* meter exist, but the basic principle of operation is the same for each. Simply stated, the basic principle of operation is to provide a continous rotating-piston motion where two or more isolation chambers deliver a discrete quantity of fluid in a portion of a cycle. As shown in Fig. 8-6, a typical *rotary-vane* meter may be comprised of an impeller driven by the fluid, a cam to move a set of vanes into the fluid stream to form an isolation chamber, and a spring to retract the vane at the end of its cycle. The space between two adjacent vanes forms the isolation chamber and the revolutions of the impeller are totalized and registered in volumetric terms.

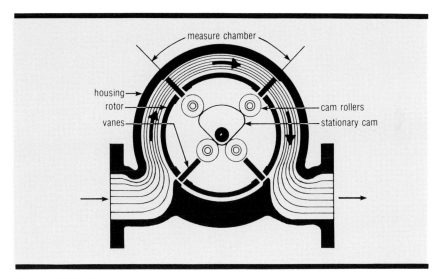

Fig. 8-6. Rotary Vane

Another form of rotating impeller-type positive-displacement meter is the *lobed impeller and gear meter*. The general features of this group of meters are shown in Fig. 8-7. Isolation chambers are formed between gears, or rotor lobes, and the outside case of the flowmeter. When two or three lobes are used on each of two rotors, equal toothed spur gears are used on the rotor shafts to keep them synchronized. Careful machining is required to keep the clearance—between the rotors and case and between individual rotors—small to insure effective capillary sealing of the isolation chamber.

Normally, all types of positive-displacement flowmeters are calibrated or "proved" to insure a high degree of accuracy. Requirements for accuracy depend on many factors, such as meter size, type of service, legal or contract requirements, and local government or company practices. For example, in the case of gasoline-dispensing flowmeters, most state and local governments require an accuracy of ±1% for a newly installed flowmeter; however, before any adjustments are required, due

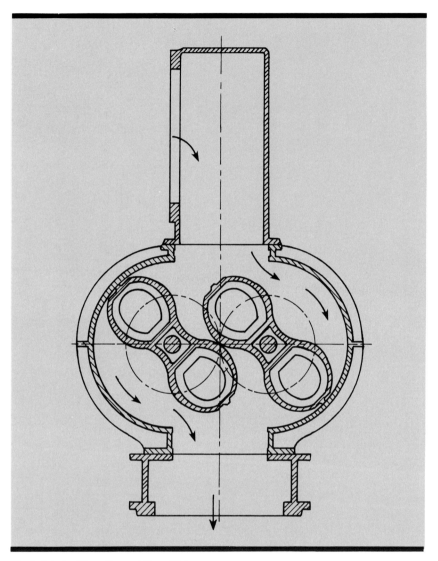

Fig. 8-7. Lobed Impeller and Gear Meter

to flowmeter inaccuracies, a ±2% error must be exceeded. With proper care and calibration, however, most positive-displacement (PD) flowmeters may be considered to be better than ±1/2% of rate total accuracy.

8-2. Current-Type Flowmeters

The difference in form between the positive-displacement-type and the current-type pulse-class flowmeters is slight, the greatest difference being that the current type does not physically capture a discrete volume of fluid but rather infers the total quantity of flow from the reaction of the fluid on the turbine or wheel-like propeller. When used in pipes, current-type flowmeters exhibit a great deal of slip, several percent as compared to the very small slippage, 0.2%, found in positive-displacement flowmeters. Greater clearances between the rotor and housing mean lower susceptibility to jamming and damage caused by solids in the flow and, more important, rotor stoppage does not cause total blockage of the flow as in the case of the positive-displacement flowmeters.

Another advantage of the current-type flowmeters is their ability to operate in large, open channels or in the open atmosphere. It is this specific advantage that allows these devices to fall into the "current" type of flowmeter category, since the earliest forms were applied to the measurement of river currents. Generally, the designs consist of a propeller, cups, or windmill-like vanes which are pushed or rotated by water or air currents.

A typical *propeller*-design current flowmeter is shown in Fig. 8-8. Commonly used for open-channel water measurement, the device has two or three helical-shaped blades mounted on a horizontal shaft. Blade diameters may range from one to five inches in diameter and flow rates from 0.1 to 15 feet per second may be covered with this type device.

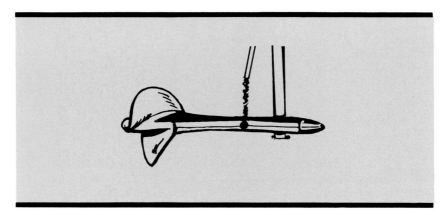

Fig. 8-8. Propeller-Type Current Meter

A *cup*-design current-type flowmeter, commonly used for open-channel measurement, shown in Fig. 8-9, is known as the Price meter. As many as five or six conical cups form a wheel and the range normally covered by a meter with a five-inch diameter wheel is about 0.1 to 15 feet per second.

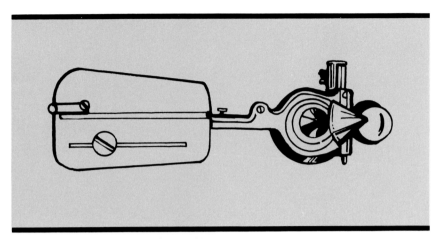

Fig. 8-9. A Five-Arm Cup-Type Current Meter

It is general practice to calibrate each current-type instrument before use and these calibrations are normally made by towing the meter in a long channel of still water at several constant speeds.

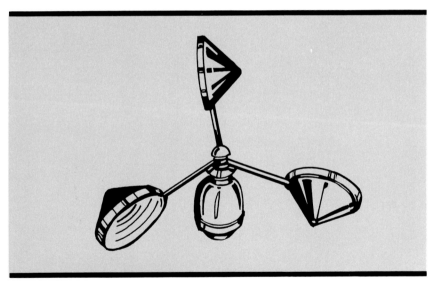

Fig. 8-10. Three-Arm Cup Anemometer

Cup-design and propeller-design flowmeters are also used to measure wind velocities. A most common instrument used in weather stations and on boats is the *cup anemometer*, shown in Fig. 8-10. The term *anemometer* is traditionally used in reference to any means for measuring the velocity of air. Cup anemometers have either hemispherical or conical cups mounted, in a three- or four-cup arrangement, to a vertical shaft. Rotor diameters may be from three to eight inches while cups range from one to five inches in diameter. A practical lower velocity measurement limit is on the order of five feet per second and the upper velocity limit is usually about 100 feet per second. Inaccuracies over this 20 to 1 flow-rate range may be as large as ±1 1/2 feet per second; therefore, considerable errors are incurred at the low velocities.

Another common mast-mounted anemometer is the *propeller* design shown in Fig. 8-11. A design of this type has one additional advantage in that wind speed as well as wind direction may be measured because the large vertical fin continually aligns the propeller in the axial flow direction.

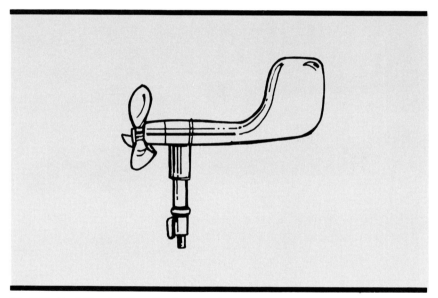

Fig. 8-11. Mast-Mounted Propeller Anemometer

The shrouded *vane anemometer* shown in Fig. 8-12 is normally employed as a hand-held device for the measurement of air flow in mine shafts, for example. Vane anemometers may range in size from less than three to over fifteen inches in outside diameter. It is important to note, that when a *vane anemometer* is used to measure flow rate in a duct, the calibration constant or meter factor may be in great error if the minimum duct size is less than ten times the outside diameter of the flowmeter.

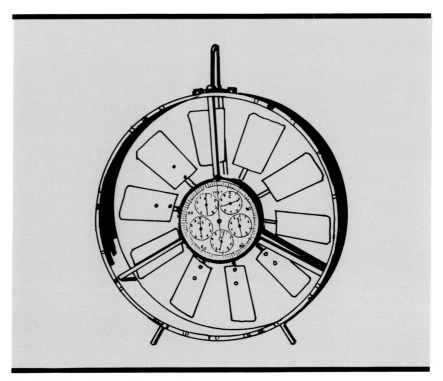

Fig. 8-12. Shrouded Vane Anemometer

This type device may be used to cover air speeds from 1/2 foot per second to 150 feet per second; however, any reasonable degree of accuracy is expected only over a 10 to 1 flow-rate range.

As for the water flowmeters, all current-type anemometers must be calibrated before use and a calibrated wind tunnel is generally used for this purpose.

Current-type meters used in a pipeline are specifically named *turbine* flowmeters to distinguish these devices from the propeller design. Most turbine flowmeter designs incorporate a meter housing with end flanges or fittings for connection to a pipeline, as shown in Fig. 8-13. A turbine-like rotor is mounted axially in the housing and the flow-rate signal is recorded by a revolution counter (register) or recognized as a pulse rate by a sensor monitoring the passage of the turbine blades. All turbine flowmeters incorporate a section of straightening vanes ahead of the rotor, to insure the fluid stream entering the rotor is free from swirl. Any velocity components in the radial direction arising from a swirl condition either speed up or retard the rotor, depending on the direction of the swirl.

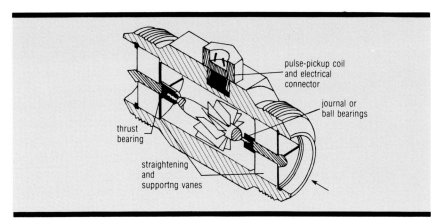

Fig. 8-13. Turbine Flowmeter

A distinction is made between turbine flowmeters used for liquid flow and those used for gas flow: gas turbine meters are characterized by a large central hub, as seen in Fig. 8-14. The large hub is used to create a venturi effect by decreasing the area and thereby increasing the velocity entering the rotor. The pressure acting on the rotor blades and hence the torque driving the rotor depend on the gas density and the square of the velocity. It is important to increase the velocity entering the rotor so that sufficient torque is available to drive a register.

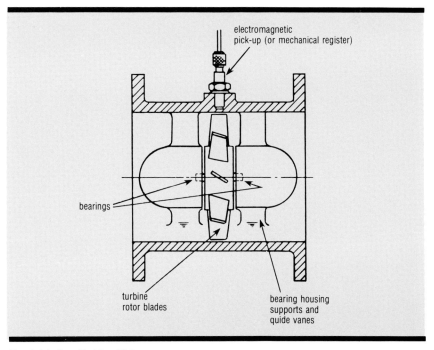

Fig. 8-14. Gas Turbine Meter

In all propeller, cup, turbine, or vane-current type flowmeters there exists a balance between the driving torque and retarding torques arising from bearing friction and the hydrodynamic drag of the rotor. Torque due to hydrodynamic drag varies as the square of the fluid velocity and bearing friction drag is essentially constant above a certain threshhold. In the design of current meters, both retarding torques are minimized; however, the effects of both are seen in the typical calibration and performance curves shown in Fig. 8-15. Note, the frequency, f_i, rotational speed of the rotor, is ideally linearly porportional to the flow rate; however, at low velocities the calibration curve A is nonlinear. At low flow rates, the bearing friction becomes more dominant and the rotor slows more quickly with decreasing flow rate thereby reducing the actual frequency f_a. Consequently, the meter factor in terms of linearity departs significantly from the mean values at the lower flow rates, usually starting at approximately two feet per second. Hydrodynamic torque variation exhibits itself as a variance in the performance in that the meter factor is not exactly constant over the linear range as demonstrated in curve B of Fig. 8-15.

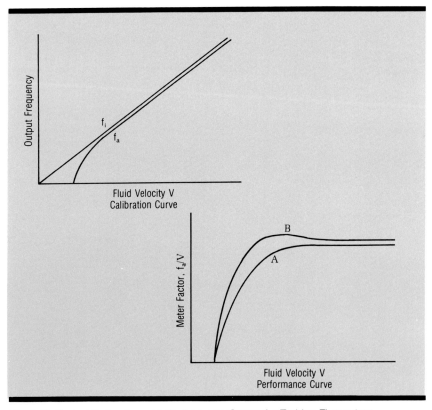

Fig. 8-15. Typical Calibration and Performance Curves for Turbine Flowmeters

Turbine flowmeters, when calibrated, are normally accurate to ±1/2% of rate over a 10 to 1 flow rate range, where the upper limit of the range may be set by any one of several considerations. For example, there is a mechanically safe upper limit of rotational speed for bearings and rotor structure. Cavitation of the blades, a condition where the pressure drop across the blade is high enough to vaporize the liquid, would limit the useful range of the meter. In addition, pressure drop across the meter may be a consideration.

Turbine flowmeters are available in sizes from 1/4 inch to 20 inches and cover flow rates from 0.7 to 30,000 gallons per minute for liquids and from about 100 standard cubic feet per minute at line pressure of 0.25 psig to 230,000 SCFM at 1400 psig line pressure.

8-3. Fluid-Dynamic-Type Flowmeters

Fluid-dynamic-type flowmeters are adolescent by flowmetering standards of growth and maturity, having only appeared on the process flowmeter market scene in the 1960s. Two basic physical phenomena are employed in the fluid-dynamic-type flowmeter. They are *vortex formation* and the *Coanda effect*. Both phenomena produce a digital or pulse output arising from the natural physics and dynamics of the fluid; hence the general category type *fluid-dynamic flowmeters*. The natural pulse output relating frequency to flowrate is produced without the use of any moving parts making these flowmeters inherently more reliable than a moving part *current-type* flowmeter.

Vortex formation devices fall into two categories, the *vortex shedding* and the *vortex precession* flowmeter.

Vortex shedding is produced by the use of a blunt, normally flat-faced, body placed perpendicular to the flowing fluid, as shown in Fig. 8-16. As fluid passes the blunt vortex generating element, the sharp corners cause a fixed point of fluid separation that forms a shear layer. In the shear layer, high-velocity gradients exist and the fluid within the layer is inherently unstable. After some length of travel, the fluid in the layer breaks down into well-formed vortices. These vortices are formed and shed with a frequency that is linearly proportional to the fluid velocity.

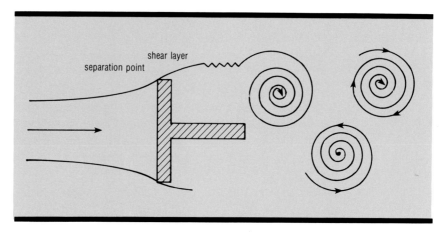

Fig. 8-16. Blunt, Flat-Faced Vortex Shedding Body

This fact had been known since 1878 when V. Strouhal observed that the frequency of eddies or vortices produced behind a bluff body increased with flowrate in a linear fashion. Although this relationship was known for many years, the measurement of the shedding frequency in a closed conduit with a sensor rugged enough to withstand process environments was a major obstacle. Nevertheless, the challenge was taken and in the early 1970s industrial flowmeters such as the one shown in Fig. 8-17 appeared on the market.

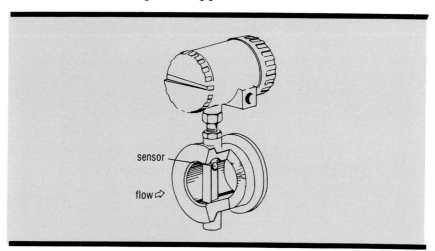

Fig. 8-17. Industrial Vortex Shedding Flowmeter

Vortex shedding flowmeters are generally comprised of three basic parts: a vortex generating element, a sensor to convert the vortex energy into an electrical pulse, and a transmitter for signal amplification, conditioning, and transmission. Differences in vortex flowmeter design lie in the shape of the element, the type of sensor, and the degree of electronic complexity required in the transmitter, which is dependent on the sensor philosophy employed. These three factors affect performance as well as cost.

The passage of vortices may be sensed in two ways: either by means sensitive to the fluctuating pressure in the wake of the bluff-body vortex generator, or by means to sense local velocity fluctuations around the body. Many different sensor concepts are employed and the sensor may be the Achilles heel of a particular manufactured flowmeter. For example, moving parts may be employed in the sensor concept, which defeats one of the attractive features of this approach to flow measurement.

Vortex-generator body shapes are important to the overall hydrodynamic performance of the flowmeter. Typical shapes employed and the sensing means associated with the shape are shown in Fig. 8-18.

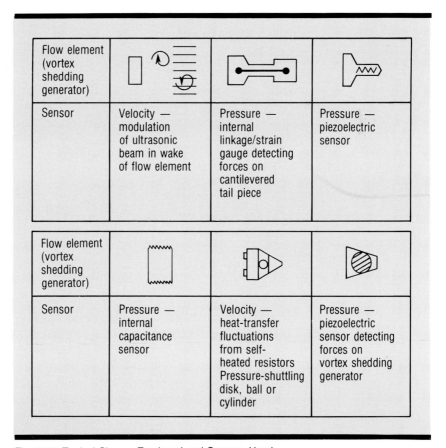

Fig. 8-18. Typical Shapes Employed and Sensors Used

Vortex shedding flowmeters generally are available in pipeline sizes from one to eight-inches. As the line size increases, the resolution (pulses per volume of fluid) decreases. This fact is easily proven from Strouhal's equation relating velocity to frequency:

$$S = \frac{fh}{V}$$

where S = Strouhal number, a constant
h = bluff body face width
f = shedding frequency
V = average flow velocity

By introducing the pipe diameter (D), one may rearrange the Strouhal equation such that:

$$S = \frac{f(h/D)D}{V}$$

Because (h/D) is a constant and independent of line size, just as the Strouhal number, the above equation shows the shedding frequency is inversely proportional to pipe diameter for a fixed fluid velocity.

$$f = \frac{KV}{D}$$

Consequently, overall accuracy may suffer for large line sizes due to the lower resolution.

The linear flow-rate range of vortex shedding flowmeters is approximately 20 to 1 for liquids and 100 to 1 for gases. In liquids, the high velocity is normally limited either by structural requirements or cavitation. In a gas, the velocity is limited only by the strong compressibility effects experienced when sonic velocities are approached.

More important than the upper velocity limit is the lower velocity limit. Most meters on the market are limited to the measurement of vortex shedding down to a velocity of approximately one foot per second only because of the limitation in the sensor. The usable lower limit, however, is Reynolds-number dependent and the lower limit of pipe Reynolds number for most meters is approximately 15,000. Below this value, the flowmeters tend to become very nonlinear and performance suffers. For very viscous fluids, the lower velocity limit is well above the one foot per second nominal threshhold of sensitivity for most sensors. Consequently, the usable range of vortex shedding flowmeters is reduced with high-viscosity fluids. A graph showing the effect of viscosity is presented in Fig. 8-19.

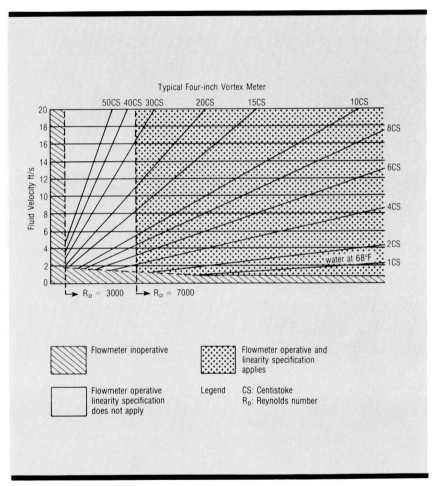

Fig. 8-19. Effect of Viscosity on Usable Range

An average linearity for most flowmeters is on the order of ±1% of flow rate over the useful range of the meter. Some manufacturers claim linearities as low as ±1/2% over a 20 to 1 range. A typical performance curve is shown in Fig. 8-20. Note the different character of the curve when compared to a turbine meter calibration. A well-made turbine meter may achieve linearities on the order of ±0.25%; however, the onset of nonlinearity occurs at a higher velocity than for the vortex flowmeter.

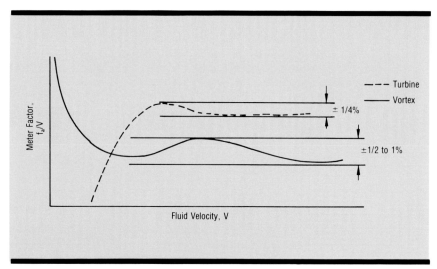

Fig. 8-20. Typical Vortex Shedding Flowmeter Performance Curve

Overall pressure loss in the vortex shedding flowmeters is about the same as for turbine flowmeters and both are approximately equivalent to the pressure loss experienced with a 0.6 to 0.7 β ratio orifice plate or 5 to 6 psi at 20 ft/sec for water flow.

Another form of vortex flowmeter is the *vortex precession* flowmeter. In operation, the fluid entering the meter is forced

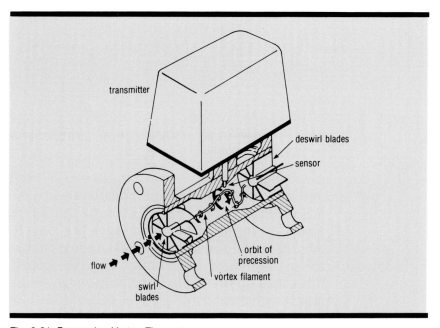

Fig. 8-21. Precessing Vortex Flowmeter

into a swirl condition along the axis of flow by guide vanes as shown in Fig. 8-21. This swirl is a vortex filament that is continously produced rather than periodically produced like the shed vortices. At the exit of the guide vanes, the flow is contracted and then expanded in a venturi-like passage, thereby causing the vortex filament to adopt a helical path. The helical path results in a precession-like motion of the vortex filament at a fixed downstream station. A sensor placed at the downstream station relays the frequency of precession which is linearly proportional to flowrate.

The precession device has overall accuracy and rangeability comparable to the vortex shedding device; however, it is available only for gas-flow application. One major disadvantage of the precession device is that the overall pressure loss is approximately five times that of the vortex shedding flowmeter.

For further reading on *vortex shedding* and *vortex precession*, consult Refs. 1 through 4.

Fluid-dynamic-type pulse-class flowmeters using the *Coanda* effect are configured as a fluidic oscillator, shown in Fig. 8-22. The geometry is arranged so that when flow is initiated, the flowing stream attaches itself to one of the two side walls. This phenomenon of fluid attachment to one wall in the presence of two walls is known as the *Coanda* effect. As shown in Fig. 8-22(A), a small portion of the flow is diverted through a feedback passage to a control port. The feedback flow, acting on

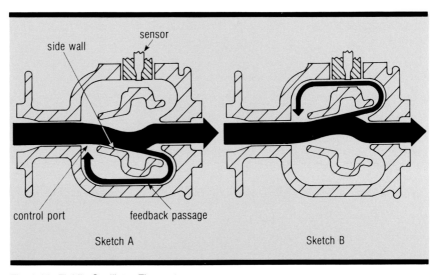

Fig. 8-22. Fluidic Oscillator Flowmeter

the main flow, diverts the main flow to the opposite side wall where the same feedback action is repeated, Fig. 8-22(B). A continuous self-induced oscillation of the flow results between the meter body side walls and the oscillation frequency is linearly proportional to fluid velocity.

As the main flow oscillates between the side walls, the velocity of the flow in the feedback passages cycles between zero and a maximum. The feedback passages thereby contain a region of substantial flow-rate change where the frequency of the oscillating fluid is readily detected. A thermal sensor may be used, for example, when a thermistor is employed as part of a constant temperature Wheatstone bridge circuit. Flow and no-flow conditions at the sensor tip induce correspondingly high- and low-heat transfer conditions between the thermistor and the fluid, which cause the thermistor to vary its resistance. The bridge circuit is designed such that its voltage automatically changes to maintain the thermistor at a constant overheat temperature. The frequency of the bridge voltage change is directly related to the oscillating flow in the meter body.

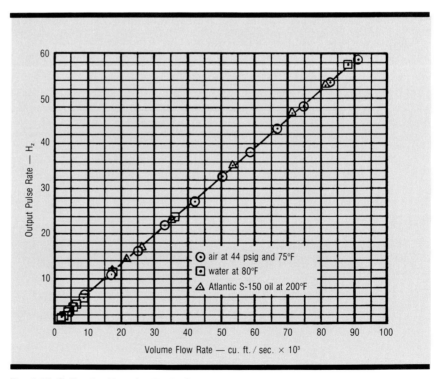

Fig. 8-23. Calibration Data for Fluidic-Oscillator Flowmeter

Like the vortex flowmeter, the fluidic flowmeter is insensitive to density and responds only to velocity changes. In Fig. 8-23, typical calibration data are shown for a one-inch meter operating with fluids of three different densities. Just as for the vortex flowmeter, any particular meter is characterized by a meter factor defined by the slope of the straight line calibration curve. Usually, the meter factor is expressed in terms of pulses per gallon or pulses per cubic foot.

In comparison to the vortex shedding flowmeter, the fluidic flowmeter is not as linear and the linearity is typically expressed a $\pm 1\%$ of full-scale flow rate. Remember, a percent of full-scale accuracy statement does not represent the same accuracy over the operating flow-rate range of the meter as does the percent of actual statement.

A minimum flow rate, below which oscillations cannot be sustained, is due to the weakening of the Coanda effect in the lower Reynolds number range. The low-velocity limit is sensitive to viscosity and the general rule of thumb for calculating the minimum flow rate is to multiply the meter body size (in inches) by the kinematic viscosity (in centistokes). The minimum flow rate is then expressed in gallons per minute.

$$\text{(inches)} \times \text{(centistokes)} = \text{GPM}$$

Irrecoverable pressure loss for the fluidic flowmeter is about the same as for the vortex devices. More information on this particular may be found in Ref. 5.

As a general note to conclude this unit, the sizing and installation of pulse-class flowmeters should be discussed. The sizing of all pulse-class devices is easy, because the manufacturer provides the information as to the maximum and minimum flow rates for a given pipeline size flowmeter. Consequently, there are no calculations required, such as for most of the head-class devices. Only the user's pipeline size and flow-rate range are needed to properly size the flowmeter.

Installation requirements for positive-displacement devices differ markedly from the current and fluid-dynamic types of flowmeters. Little concern is given to the upstream disturbances preceding positive-displacement-type flowmeters because the operating characteristic of these meters is such that they handle discrete quantities of fluid. Each fluid quantity is segmented or compartmentalized so that little concern need be given to the upstream velocity profile shape. On the other hand, current and fluid-dynamic devices are sensitive to upstream disturbances. A general rule to follow for these devices is to install them according to the manufacturer's specifications. Most manufacturers recommend following orifice-installation practices.

This concludes our discussion on the *energy extractive approach* to flow measurement. The next unit will deal with powered flowmeters which use an *energy additive approach* to flow measurement.

Exercises

8-1. Indicate to which type(s) of pulsed-class flowmeter the statement applies:

	Positive displacement	Current type	Fluid-dynamic
(a) Requires no moving parts			
(b) Sensitive to approach conditions			
(c) High % slip			
(d) Frequency of pulses linearly related to velocity			
(e) Counts discrete fluid parcels			
(f) Nutating disk			
(g) Turbine flowmeter			
(h) Cup anemometer			

Unit 8: Pulse Producing Flowmeters

8-2. The flow rate through a test loop is varied over a 20 to 1 range. Two flowmetering systems are used to monitor the flow rate. One is an orifice/differential pressure system, the other is a turbine flowmeter. What is the range of the output signal for each flowmeter?

8-3. A vortex flowmeter, $h/D = 0.3$, operates over a 20 to 1 flow-rate range with a constant Strouhal number, $S = 0.127$. The pipe diameter is four inches and the minimum rated flow rate is one fps.
(a) What is the meter factor, K, in pulses per cubic foot of flow?
(b) How many cubic feet of flow does each pulse represent?
(c) What is the vortex shedding frequency at the maximum flow rate?

8-4. Select the type of pulse-class flowmeter best suited for metering a highly viscous, non-Newtonian flow.

8-5. A flow of water and sand (~1%) must be metered. Which type of pulse-class flowmeter would retain accuracy best over a long period of operation?

References

[1] Cousins, T. and Nicholl, A.J., "Comparison of Turbine and Vortex Flowmeters," *CME*, February 1978.
[2] Dijstelbergen, H.H., "The Performance of a Swirl Flowmeter," *Journal of Physics*, 1970, 3(11)886-888.
[3] Miller, R.S., DeCarlo, J.P., and Cullen, J.T., "A Vortex Flowmeter—Calibration Results and Application Experiences," NBS Special Publication 484, Vol. 2, pp. 549-570.
[4] Inkley, F.A., Walden, D.C., and Scott, D.J., *Flow Characteristics of Vortex Shedding Flowmeters, Measurement and Control*, Vol. 13, May 1980.
[5] Adams, R.B., *A Fluidic Flowmeter*, ISA Reprint 73-815.

Unit 9:
Powered Flowmeters

Unit 9

Powered Flowmeters

Rather than extract energy from the flowing fluid to monitor flow rate, one may add energy to the flow and monitor either the effect of the flowing fluid on the introduced energy source or the effect of the energy source on the fluid. Remember, this approach to flow measurement is called the *energy additive approach*.

Three classes of flowmeters using the *energy additive* approach to flow measurement are the *magnetic class*, the *sonic class*, and the *thermal class*. Each class relies on electrical power to introduce the energy source into the flow and the three energy sources are magnetism, sound, and heat.

Learning Objectives—When you have completed this unit you should:

A. Understand the principle of operation of each of the three classes of powered flowmeters.

B. Know the different approaches taken, utilizing the same power source, within each class.

C. Be familiar with the accuracy, use, and application of each class of powered flowmeter.

9-1. Magnetic Flowmeters

In 1831, British physicist Michael Faraday measured a voltage induced in a circuit by imposing a relative motion between the circuit and a magnet. Shortly thereafter, Emile Lenz found the direction of the induced current was such as to create a field opposing the relative motion that caused it.

These fundamental discoveries represent the technical foundation for the electromagnetic flowmeter and, in fact, Faraday attempted to measure the flow of the Thames River using the earth's magnetic field and electrodes placed on opposite banks of the river. His attempt was doomed to failure because of polarization of the electrodes. The reason for the polarization will be explained later. Because of this

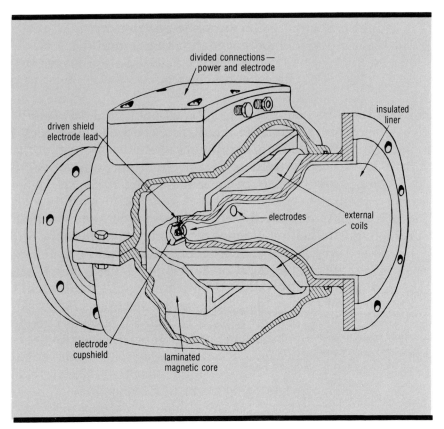

Fig. 9-2. A Process Magnetic Flowmeter

Power necessary to produce the alternating magnetic field is usually supplied from local electrical supply lines. These lines are not closely controlled in amplitude, frequency, and waveform, therefore, the output signal of the flowmeter is affected by variations in the magnetic field, B, caused by power-line variations. Consequently, it is necessary to ratio the output signal to a line-reference signal in order to obtain the true flow rate. Two common reference signals in use today are the current in the magnetic coils or the voltage supply to the coils. Other reference signals may be derived from a coil wound around the iron of the magnet or from a coil placed in the air gap of the magnet assembly.

Other factors affecting the output signal are the phase relation between the flow signal, e, and the reference signal, and that changes in the fluid conductivity may affect the phase of the output signal. Also, stray signals may be generated by the leads from the electrodes by the varying flux intersecting the leads which, in effect, form a coil.

The secondary device, that is, the electronics in the signal-processing part of the flowmeter, is designed to eliminate or account for all the above signal abberations. The primary device is comprised of the pipe or flow tube, the electrodes, and the coils and core of the electromagnet.

Pulsed Direct Current Type

Because the major source of error in the AC electromagnetic flowmeter is due to stray currents causing zero shifts and the resulting need to physically bring the fluid to zero flow rate to correct for baseline or zero conditions, instrument manufacturers looked to other schemes for magnetic flowmeter signal generation.

A relatively new and attractive scheme is the *pulsed DC magnetic flowmeter*, Ref. 2. Basically, a direct current is periodically applied to the magnetic coils, thereby avoiding any residual DC voltages on the electrodes arising from galvanic and thermal action. The DC output signal is proportional to the velocity, as in the case of the constant magnetic field flowmeter, however, polarization is avoided. In operation, the DC output frequency signal is detected and stored during the excitation time of the magnets and when the magnets are in the unexcited state. Any signal noise present in the time period when the magnets are not excited is due to extraneous sources and is not a flow signal. Therefore, by subtracting the voltage signal stored in the unexcited state from the voltage signal in the excited state of the magnet, a signal proportional only to the flow rate results. A schematic representation of this signal-sampling scheme is shown in Fig. 9-3. As seen in the sketch, no measurement is taken in the time period of magnetic-coil excitation until the signal level is stable and of constant value. After the coil is deenergized and a stable signal is attained, the two signals are subtracted and output as the flow signal.

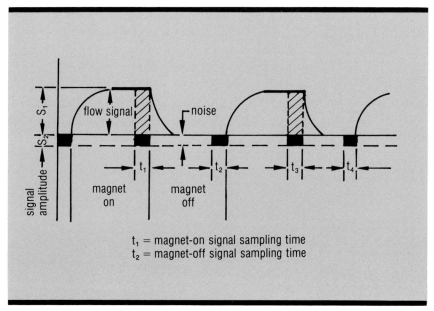

Fig. 9-3. Pulsed DC Signal-Sampling Scheme

Consequently, some of the apparant advantages of the pulsed-DC type over the AC type are that stray signals are effectively ignored and a stable zero or baseline reference results, and the pulsed-DC approach requires less power and resultant lower power consumption.

Characteristics of Magnetic Flowmeters

Magnetic flowmeters will measure only the flow rate of conductive fluids, however, the threshold conductivity for many meters on the market is very low, on the order of 0.1 (microsiemens/cm). Most petroleum products (as well as gases, except if the gas is highly ionized) fall well below the practical limits of conductivity.

Magnetic flowmeters measure the velocity of the flow, as seen in the basic equation. If the magnetic flux field is uniform across the flow tube, an integrated velocity is measured. Consequently, if there are velocity profile changes due to swirls or helical flow patterns, for example, the total measured velocity is unaffected as long as the velocity profile across the pipe is symmetrical. Nonsymmetric flow profiles may cause flow-rate measurement errors on the order of several percent.

As a result of the ability of magnetic flowmeters to integrate at least symmetric velocity profiles, the flowmeter is often considered a volume rate flowmeter. The flowmeter is insensitive to density and viscosity, can measure flow in both directions and, being obstructionless, is often used in slurry flows. For example, in the mining industry, the flow rate of metal-bearing abrasive slurries such as copper ore are measured. Other applications cover a wide range of fluids from waste water, sludge, paper stock, bleaches, dyes, acids, and emulsions, to beer, soda, and milk.

If the primary and secondary components of a magnetic flowmeter are flow calibrated as a unit, system accuracies as low as $\pm 1/2\%$ of rate are achievable. However, normal accuracy specifications on both the AC and pulsed-DC devices are $\pm 1\%$ of actual flow rate.

9-2. Ultrasonic Flowmeters

In the *sonic class* of flowmeters, two important types will be discussed: the *time-of-flight (TOF)* and the *Doppler* ultrasonic flowmeter.

For both types, electrical energy is used to excite a piezoelectric crystal type of material to a state of mechanical resonance. The crystal is either placed in contact with the fluid (*wetted transducers*) or mounted on the outside of the conduit containing the fluid (*clamp-on transducers*). As the crystal resonates, a sound wave traveling at the speed of sound of the media is generated and this sound wave is used to interrogate the flow field for the purpose of extracting the flow rate.

Time-of-Flight Type

One of the first inventions of an acoustic contrapropagating transmission (upstream-downstream) flow measurement apparatus for use in a pipe was patented in Germany by a man named Rütten in 1928. In 1954, H. P. Kalmus measured flow velocity in a pipe using externally mounted transducers which he used for generating and detecting contrapropagating waves. However, it wasn't until the late 1970s that the ultrasonic flowmeter was introduced.

In the *time-of-flight (TOF)* type of *sonic-class* flowmeters, a sound wave is introduced to the flowing fluid in such a way that the sound wave alternately travels against the flow in one direction and with the flow in the other direction, as shown in Fig. 9-4. The difference in transit time of the wave is proportional to the fluid flow rate because the sound wave is slowed when traveling against the fluid flow and accelerated when traveling with the flow. Think of rowing a boat (sound wave) the same distance upstream and downstream in a flowing river (pipe flow). The difference in the time it takes to row upstream as opposed to rowing downstream is proportional to both the velocity of the river and the velocity of the boat. If one could keep the velocity of the boat constant and known relative to a fluid particle in the river, then the velocity of the river could be calculated (see Fig. 9-4). Equivalently, if the sound wave velocity of a fluid is known as well as the time difference and distance traveled, the fluid flow rate is known.

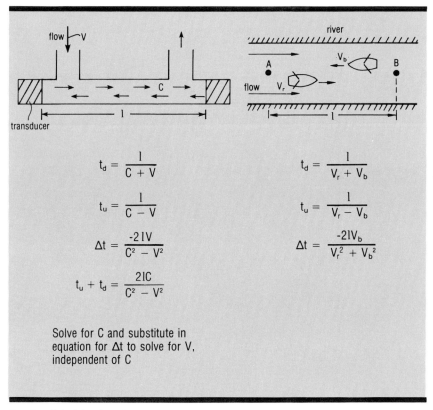

Fig. 9-4. Principle of the Ultrasonic Flowmeter

Unfortunately, in the process world, the speed of sound in a process fluid is usually not only unknown, but variable with temperature as well. Compensation for speed-of-sound changes may be accomplished by utilizing the sum of the transit times in conjunction with the time difference to develop a signal independent of sound speed. Another approach taken is to eliminate the effect of the speed of sound by monitoring the sound-wave frequency difference. A series of sonic pulses of known frequency traveling upstream is subtracted from a similar series of sonic pulses traveling downstream. The measured difference between the frequencies is a direct function of the flow velocity and independent of the velocity of sound.

Consider the following equation while keeping in mind the development and structure shown in Fig. 9-4.

$$f_d = \frac{C + V}{l}$$

$$f_u = \frac{C - V}{l}$$

$$f_d - f_u = \frac{2V}{l}$$

Because the frequency is essentially the inverse of the time, the sound velocity is cancelled, whereas, in the time domain, the square of the sound speed dominates.

Consequently, time-of-flight ultrasonic flowmeters may use some form of continuous frequency difference scheme rather than time difference with temperature compensation. An example of one is the up-down counter. The sound-wave pulse frequency traveling upstream is counted for a period of time and is subtracted from a similar time sample for the downstream-traveling pulse frequency. A pulse frequency is typically generated by a *sing-around technique* where a received pulse triggers a transmission pulse.

Several forms of the time-of-flight (TOF) ultrasonic flowmeter are shown in Fig. 9-5, namely the axial transmission, multibeam, single-beam sing-around, reflected beam, and cross-beam forms.

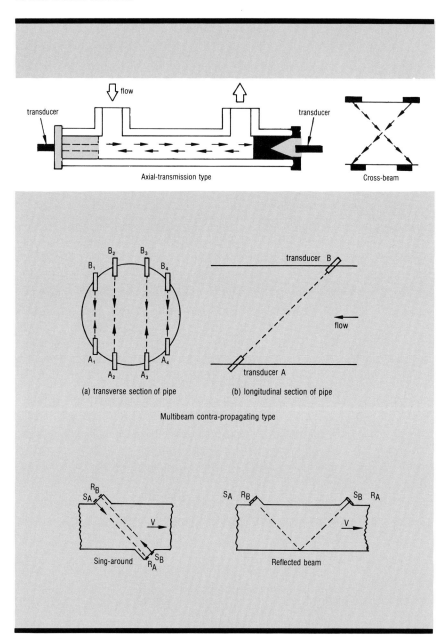

Fig. 9-5. Several Forms of Time-of-Flight (TOF) Ultrasonic Flowmeters

Generally, the wetted transducer devices are considered to be more accurate than the clamp-on flowmeters because of two factors, that is, the smaller angle of the sonic beam relative to the flow direction and lower transmission losses due to the absence of the pipe wall. In the wetted tranducer arrangement, Fig. 9-6, a 45° transmission angle is normally chosen to save on pipe length while optimizing the amplitude of the velocity vector along the sound path. The vector of flow along the sound path is the cosine of 45° times the flow velocity. For the clamp-on arrangement, Fig. 9-6, the angle of transmission is considerably less than for the wetted transducers when steel is the pipe wall material. For example, the transmission angle may be on the order of 60° or 75°, depending on the sound-refraction angle between the steel and the liquid. The cosine of 75° is 0.259 as compared to 0.707 for the 45° transmission angle; therefore, the amplitude of the velocity vector along the transmission path is greater for the wetted transducers than for the clamp-on transducers. All things being equal, the wetted transducers should deliver a higher-quality signal and potentially the more accurate flow-rate measurement.

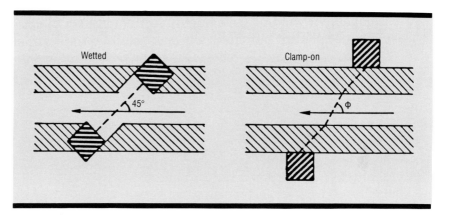

Fig. 9-6. Wetted and Clamp-on Transducer Arrangement

Doppler Type

The Doppler effect discovered in 1842 may be experienced by listening to a horn-blowing, oncoming car or train as it passes by. The tonal quality (frequency) heard as the train approaches is different than that heard as the train departs. When an ultrasonic beam is projected into an inhomogeneous fluid, some acoustic energy is scattered back toward the transducer. Because the fluid is in motion relative to the fixed transducer the scattered sound moving with the fluid is received by the transducer at a different frequency than it was sent. This different frequency is known as the Doppler-shift frequency and the difference between the send and received frequencies is directly proportional to the fluid flow rate.

Accuracy specifications for ultrasonic flowmeters vary significantly according to manufacturer and type of flowmeter. In general, wetted transducer TOF devices are the most accurate and, calibrated under laboratory conditions, may deliver ±1% of rate performance. However, a more realistic specification is on the order of ±1/2% of full scale. The clamp-on Doppler devices are at the other end of the accuracy scale and, in general, not more than ±1% of full scale can be expected.

Because ultrasonic flowmeters are new relative to such flowmeter standards as the orifice plate or venturi, there is a great deal of evolution at present in the technology. With the advent of sophisticated electronics and greater application experiences, no doubt overall performance in these devices will improve in the future. For additional reading material on these devices see Refs. 3, 4, and 5.

9-3. Thermal Flowmeters

In the thermal class of flowmeter, flow rate is measured either by monitoring the cooling action of the flow on a heated body placed in the flow or by the transfer of heat energy between two points along the flow path. Two types of flow-measurement devices in the thermal class of flowmeter are the *thermo-anemometer* and the *calorimetric flowmeter*, see Refs. 6 and 7.

Origins of the thermal class of flowmeter can be traced to the last two decades of the nineteenth century through the work of L. Weber, 1894, and Oberbeck, 1895. Although the hot-wire anemometer was invented by Weber, it was developed for use in air streams by L. V. King and applied in water by G. Gangadharan in 1931. Modern techniques were adapted by the Iowa Institute for Hydraulic Research, where S. C. Ling constructed a hot-film anemometer in 1955 and Ph. G. Hubbard developed a new hot-wire anemometer, based on a constant, relatively low temperature and an alternating electric current, in 1957. These delicate instruments were only used in laboratories.

Thermo-Anemometers

The devices used in thermo-anemometry may be divided into two groups. First are those devices in which the thermo-element is a fine metal wire on the order of 0.00015 inches diameter, as shown in Fig. 9-9, and the element is connected as one of the arms of a balanced measuring bridge. These anemometers may be made with their thermo-elements at either constant or variable temperatures. Rate of flow is measured by the variation in the element resistance for a supplied current of constant magnitude, or by the variation in the magnitude of the current for an element with its resistance, that is temperature, held constant. These devices are commonly referred to as *hot-wire* probes.

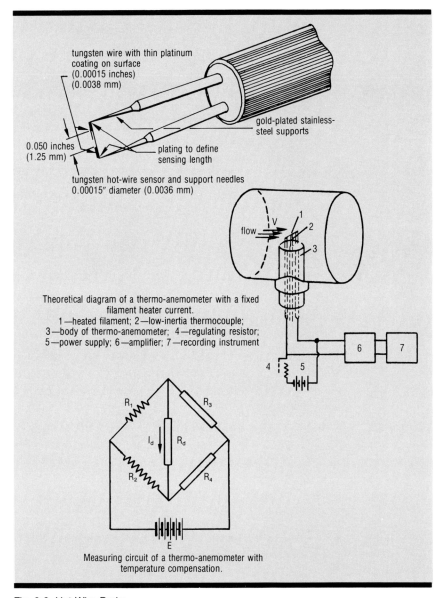

Fig. 9-9. Hot-Wire Probe

The second group of devices in the thermo-anemometer-type instrument consists of metallic films deposited on a vitreous or ceramic substrate. Probes of this design are referred to as *hot-film* probes. As shown in Fig. 9-10, these probes may be designed in various shapes: a wire typically 0.002 inches in diameter, a wedge, a cone, and flat surface shapes.

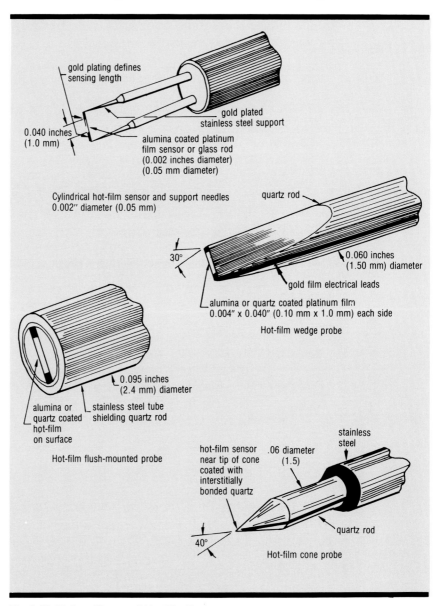

Fig. 9-10. Various Shapes of Hot-Film Probes

When compared to the *hot-wire*, the *hot-film* probe has the following advantages: it is more rugged, less susceptible to accumulation of foreign materials, and easier to clean; it has better frequency response over a wider frequency range than the hot-wire; and it is easily adapted to a variety of probe shapes.

Although *hot-film* sensors have inherent advantages over the *hot-wire* type, in some applications the *hot-wire* probe is superior. For example, because the dimensions of the hot-wire are considerably less than the film-coated wire, the hot wire is superior for measurements taken close to a wall.

Consider now the amount of heat given up (q) by a heated sensor to the flow in terms of the current supplied, I, and the wire resistance, R.

$$q = 0.24 \, I^2 R$$

In terms of the fluid properties and fluid velocity, the above equation may be expressed as:

$$q = 0.24 \, I^2 R = (t_s - t_g)\left[C_t + (2\pi d C_v \rho V)^{1/n}\right]$$

where

t_s = sensor operating temperature
t_g = fluid temperature
C_t = thermal conductivity of the fluid
C_v = thermal capacity (specific heat at constant volume for a gas)
ρ = density of the fluid
d = diameter of the wire
V = velocity of the fluid
n = usually close to 2

From the equation one can extract two important facts, namely: (1) hot-wire and hot-film probes are inherently mass flow (ρV) sensitive; and (2) this approach to flow measurement is highly fluid-composition dependent due to the presence of C_t and C_v in the equation. In addition, the output of the probe in terms of current or voltage, as shown in Fig. 9-11, is extremely nonlinear and a linearizing circuit is normally employed.

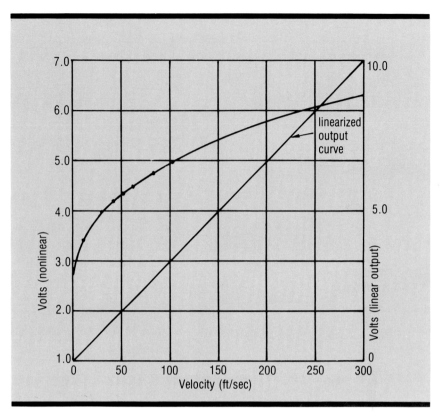

Fig. 9-11. Nonlinear Output Characteristics of Thermo-Anemometer

Most often, a constant-temperature circuit is used rather than the constant current because of the superior performance of the constant temperature approach in both noise level and frequency response. In addition, constant-temperature circuitry is compatible with the complex frequency characteristics of hot-films, whereas the constant current system is not. Operation at constant temperature increases probe life and prevents sensor burnout due to the velocity changes rapidly passing the probe causing a rapid cooling effect. This is especially important in liquids in which large sensor cooling occurs with velocity changes. Linearizing is not possible with a constant-current system, making the constant-temperature approach more attractive. Constant-temperature probes are easily temperature compensated. For these reasons, the constant-temperature approach is favored over the constant-current approach.

Calorimetric Flowmeters

Calorimetric flowmeters work on the principle of heat transfer by the flow of fluid and may be divided into three groups:

Devices drawing constant power to the heater with simultaneous measurement of the amount of heat transferred to the flow, which is velocity dependent.

Devices that heat the flow to a constant temperature with simultaneous measurement of the energy supplied to the heater, which is velocity dependent.

Devices that vary the heater temperature sinusoidally with time, where the flow rate is measured by the signal phase-shift at the sensor compared to the input signal at the heater.

In general, calorimetric flowmeters are comprised of elements arranged consecutively along the direction of motion of the flow to be measured, Fig. 9-12: a device to measure the temperature of the flow ahead of the heater, a heater, and a measuring device for the temperature of the flow downstream of the heater. The rate of flow is determined by the difference in the two temperature readings. Calorimetric flowmeters are often referred to as "heated grid" flowmeters.

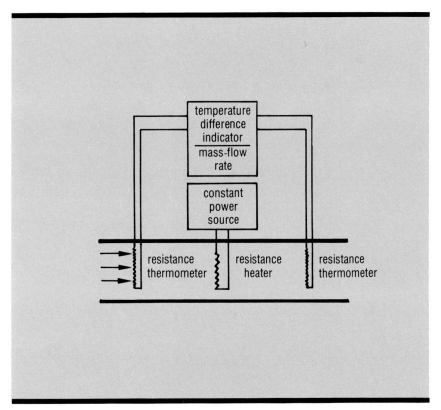

Fig. 9-12. Calorimetric Flowmeter (Heated Grid)

If the absorption of heat by anything other than the flowing fluid, for example, the pipe walls, is considered negligible, the equation for thermal balance has the form:

$$q = \rho V C_p (t_b - t_a)$$

where

C_p = thermal capacity (specific heat at constant pressure for a gas)
t_b = temperature of the fluid before the heater
t_a = temperature of the fluid after the heater

Since the heat is transferred to the flow by an electrical heater, q may be expressed in terms of current and voltage as:

$$q = \frac{K e^2}{R} = eVC_p(t_b = t_a)$$

where

e = voltage drop for a supplied current
R = resistance of the heater
K = thermal equivalent of electrical energy

Examination of the above equation reveals the fact that the calorimetric type of flowmeter is mass-flow (ρV) dependent, as is the thermal type; however, the mass-flow rate is linearly proportional to the heat transfer, rather than square law as is the thermal type.

Two other calorimetric flowmeter devices are shown in Fig. 9-13: the externally heated and the sinusoidally heated designs. In the first design, flow rate is measured by the transfer of heat by the boundary layer or flow near the wall of the tube. Although the boundary layer flow rate is not the same as the main flow, this measurement may be used to infer the main flow rate through a constant of proportionality. The dynamic response of this approach is poor because of the large thermal inertia of the walls and the time-lag introduced by the boundary layer reacting to the change in the main flow.

In the sinusoidally heated element device, a sine-wave voltage of predetermined frequency is supplied to a heater and a varying temperature field moving with the flow is generated. A

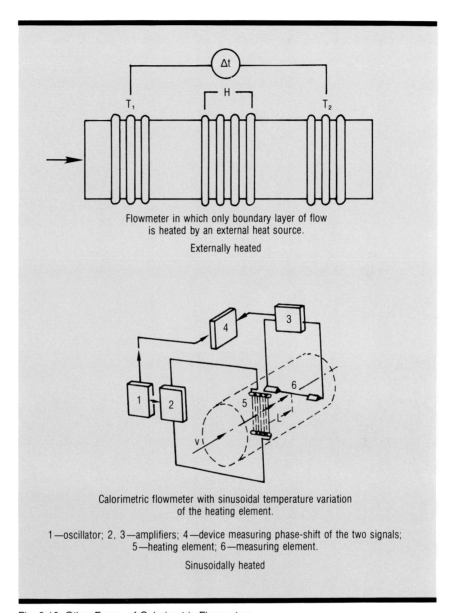

Fig. 9-13. Other Forms of Calorimetric Flowmeters

measuring device placed at a fixed distance downstream from the heater detects the temperature variation and the varying resistance of the device results in an electrical frequency signal which is phase-shifted relative to the sine-wave signal applied to the heater. This phase shift is linearly and inversely proportional to flow rate in the form:

$$\Delta \phi = \frac{2\pi fL}{V}$$

where

ϕ = phase
f = frequency
V = velocity
L = distance between heater and sensor

Note: this approach to flow measurement using heat is not mass-flow (ρV) dependent, as are the other approaches. The reason for this difference is that the basic principle is not heat-transfer dependent. The basic principle of operation is called *tagging*, which is to be discussed further in the next unit, Special Techniques.

The overall accuracy of calorimetric flowmeters is on the order of $\pm 1\%$ to $\pm 2\%$ of full scale, whereas the thermo-anemometer may be considered accurate to $\pm 1\ 1/2\%$ to $\pm 2\%$ of reading or $\pm 1/2\%$ of full scale. In any case, each probe or flowmeter should be calibrated before use.

Exercises

9-1. Why are constant-field magnetic flowmeters impractical?

9-2. What are the advantages of a pulsed-DC type versus an AC-type magnetic flowmeter?

9-3. Explain the differences between Doppler flow-measurement techniques and the "time-of-flight" technique.

9-4. In sonic-type flow measurements, which is more desirable, high or low frequencies? Why?

9-5. What are some advantages of hot-film probes as compared to hot-wire probes?

9-6. In both types of thermal flowmeters the flow rate is inferred through a thermal-energy balance on a control volume. Identify the control volume being considered for each type of thermal flowmeter.

References

[1] Kolin, A., "An Alternating Field Induction Flowmeter of High Sensitivity," *Review of Scientific Instruments*, Vol. 16, pg. 109, May 1945.

[2] Fath, J. P., "Magmeters with Ultra-Stable Zero," ISA 75-830, 1975.

[3] Powell, D. J., "Ultrasonic Flowmeters' Basic Designs, Operation, and Application Criteria," *Plant Engineering*, Vol. 33, No. 9, pg. 93, May 1879.

[4] Schmidt, T. R., "What You Should Know about Clamp-on Ultrasonic Flowmeters," *InTech* 0192-303X/81/05/059/04/, May 1981.

[5] Waller, J. M., "Guidelines for Applying Doppler Acoustic Flowmeters," *InTech* 0192-303X/80/10/0055/03, October 1980.

[6] Freymuth, P. "Review: A Bibliography of Thermal Anemometry," Transactions of the ASME, *Journal of Fluids Engineering*, Vol. 102, pg. 152, June 1980.

[7] Katys, G. P., *Continuous Measurement of Unsteady Flow*, a Pergamon Press Book, Macmillan Co., 1964.

Unit 10:
Special Techniques

Unit 10

Special Techniques

In certain harsh process environments or for reasons of cost, special techniques are used to extract a flow-rate measurement. These techniques usually employ either an *energy extractive* approach or an *energy additive* approach to flow measurement, or a combination of the two in some unique form. Three special techniques called *tagging*, *deflection*, and *by-pass* are discussed in this unit in terms of the principle of operation, application, and performance.

Learning Objectives—When you have completed this unit you should:

   A. Be familiar with the various special techniques usually applied to special flow metering needs.

   B. Understand the principle of operation of each of the special techniques.

   C. Know the performance capabilities of available instruments that employ these special techniques.

10-1. Tagging (Flow Marker or Tracer) Technique

The flow-measurement technique called *tagging* often is referred to as the *tracer* method or *flow-marker* technique. In practice, either a substance introduced into a flowing fluid or a natural marker in the fluid is traced between two points. Knowing the distance between the two points and measuring the time it takes the marker to traverse the known distance results in a direct measurement of flow rate, as illustrated in Fig. 10-1.

Tagging techniques are classified in two general categories: one where the marker is introduced into the flow by *injection* or *induction*, and the other where the marker *exists* in the flow. The difference between *injection* and *induction* is that the injection procedure requires an entry mechanism to the fluid, whereas using induction means the pipe wall or fluid boundary is not penetrated. Some examples of *induction*-type markers are: ionic-radioactive and magnetically induced tracers. Typical

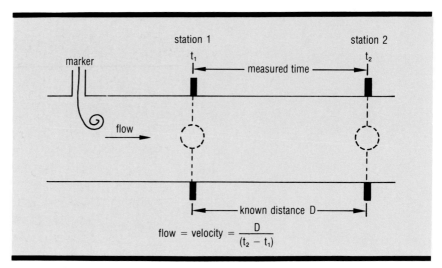

Fig. 10-1. Tagging Technique Principle

injection-type markers are salt solutions, hot or cold fluids, gases, and either flourescent or opaque dyes.

Markers that *exist* in a fluid are usually either natural dust or dirt in fluids or natural turbulence in the form of vortices or eddies. A device called a Laser Doppler Anemometer (LDA) or sometimes called a Laser Doppler Velocimeter (LDV) is used to trace natural dust or dirt. A *correlation* flowmeter is used to trace natural eddies in the flow. Both the (LDA) and *correlation* flowmeter are discussed in greater detail later in this unit.

First, *injection* and *induction*-type *tagging techniques* are discussed.

With reference to Table 10-1, note the *injection*-flow markers are listed as salt, heat, gases, dyes, and radioactive materials. The use of salt as a marker is commonly called the *salt velocity* method and sometimes the *Allen salt* velocity method. This approach to flow measurement is used only in water flows and utilizes the high electrical conductivity in the salt-water solution as the tag. In practice, a concentrated salt-water solution is injected into a water stream through a series of nozzles so that a uniform distribution of solution over the cross-section area is achieved. Downstream, an array of electrodes is placed so that equal segments of the cross-section area are covered and these electrodes react to the increase in conductivity as the brine solution passes. Either one set or two sets of electrodes array may be used. If one set is used, the time of injection must be recorded as well as the time of passage. If two sets of electrodes are used, the time of passage between the two sets is recorded.

Marker Type	Marker Substance	Applicable Flows	Detector
Conductive	Salt	Water	Electrodes
Thermal	Hot or cold fluids	Liquids or gases	Hot-wire Hot-film
Gaseous	Carbon dioxide Nitrous oxide	Gases	Infrared spectrometer
Dye	Rhodamin, Pontasil Methyl blue, sodium dichromate	Transparent liquids and gases	Flourimeter Colorimeter
Radioactive	Radiactive isotopes	Gases and liquids	Geiger counter Scintillation counter

Table 10-1. Injection Methods

Between the injection point and the first electrode section, a minimum distance of four diameters should be maintained to assure adequate mixing before the electrode array is reached. If two electrode arrays are used, four diameters between arrays are necessary to assure clean passage of the solution past the first array before reaching the second array.

A thermal-type flow marker may simply be a small quantity of fluid heated or cooled to a temperature considerably different from the process fluid temperature. Preferably, the heated or cooled fluid is the process fluid so that the process is not contaminated by the tracer. Process contamination is one of the major drawbacks in using *injection-tagging* techniques.

To detect the passage of a fluid slug having a different temperature, hot-wire or hot-film type probes often are used. Remember, the last thermal-type flowmeter discussed in the previous unit does not operate on the same basic principles as other thermo-anemometers. It operates using a *tagging* technique—one element upstream heats the flow and the other downstream records the passage of the heated fluid.

For thermal as well as most other tagging techniques, a closed-loop system may be used, where markers are formed at a frequency proportional to flow by feeding back the sensed-signal to trigger the send-signal. Also, a system may be used where flow markers are formed at a constant frequency and the time interval between sender and sensor is monitored.

The use of gas as a tracer material is usually limited by the degree to which the tracer is considered a contaminant to the process. Normally, gases having high infrared absorption characteristics are used because of the ability of infrared detectors to sense the presence of small quantities of trace material on the order of 10 parts per million; however, other gases, those that are opaque or those that radiate also are used in conjunction with optical sensors. Opaque tracers may employ simple photo-diode sensors, for example. Radiating tracers would employ a photoelectric detector sensitive to a specific narrow spectrum equivalent to that of the radiating gas. Use of gas tracers is usually limited to application on gas flows. Accuracies on the order of $\pm 1/2\%$ to $\pm 1\%$ are possible using this method; however, distances on the order of 100 feet between the injection station and the sensor and 300 feet between sensors may be required to achieve high degrees of accuracy.

Dyes often are used as marker substances, especially when optic-sensing techniques are employed. For example, flourescent materials such as Rhodamin and Pontasil may be detected using a flourimeter, especially in liquid-flow application having little or no flourescent properties. Materials such as Methyl blue and sodium dichromate are used as markers in conjunction with color sensors or a color-reagent comparison standard.

Radioactive markers are substances such as xenon, krypton, and radon injected into gas flows or powdered isotopes injected into liquid flows. Downstream of the injection point, radioactivity monitors such as Geiger or scintillation counters record the passage of the radioactive substance.

The use of *induction* methods (Table 10-2) to introduce a marker to the process have inherent advantages over the injection method in that first, no contaminant is added to the process, and second, in most cases the process line does not have to be penetrated.

Marker Type	Marker Substance	Applicable Flows	Detector
Ionic	Ionized gas	Gases	Ion collector
Magnetic	Magnetized nuclei	Liquids	Coil

Table 10-2. Induction Methods

One approach readily applicable to gas flow is to ionize the process gas by either a spark or ionize the gas through the use of a radioactive substance operating in a pulsed mode. The ionized marker is then detected by an ionization counter.

A system using modulation radioactive radiations to ionize the gas employs a radioactive isotope placed on the side of the gas pipeline. Radiation is modulated by either mechanical or electromagnetic means, so that the gas is ionized in short, alternating pulses. The ionized gas markers are detected at some distance downstream of the ionizer by two plates attached to and insulated from the walls of the pipe. When voltage is applied to these plates and the ionized marker is passed between the plates, a voltage pulse is produced.

This method is severely limited by the short duration of ionized molecules. As the conductivity or the temperature of the process increase, the life of the ionized marker decreases due to recombination of the ions. Consequently, high radiation energies and short distances between ionizer and sensor may be required.

Nuclear magnetic resonance (NMR) flowmeters utilize a tagging technique by which the marker's fluid nuclei are magnetized and subsequently traced between two stations a known distance apart. In operation, fluid first enters a magnetizer section where some degree of nuclear magnetism is imposed. After leaving the magnetizer, an inverter often is used at the first station to alternately invert the nuclear magnetic field, thereby generating a sinusoidal tag. The markers then proceed toward the second station a fixed distance downstream of the inverter. At this station, a receiver coil delivers an induced voltage proportional to the magnitude of nuclei magnetization, and a frequency proportional to the flow rate.

The use of *existing* tags or markers in a fluid stream is an attractive approach to flow measurement when contamination is a key issue. One device utilizing very small amounts of existing particles in either a gas or liquid stream is the Laser Doppler Anemometer (LDA), a point-velocity measurement device normally employed in laboratory situations rather than in the process industry. The device shown schematically in Fig. 10-2 operates on the basic principle of the Doppler effect, which is discussed in great detail in Unit 9, with regard to sonic-powered Doppler flowmeters. In operation, a single laser beam is split into two beams which are subsequently made to converge at the desired point of measurement and have identical angles and the same frequency. Natural particles, such as dust, dirt, etc., moving with the fluid velocity pass through the beam intersection point and reflect or scatter light from the beams. As the particles scatter light from both beams, the scattered light from each beam is of a slightly different frequency because the particle velocity causes a slightly higher frequency in the scattered light from the beam it moves toward and a slightly lower frequency in the scattered light from the beam it moves away from. The different frequencies in the scattered light interfere constructively and destructively forming a third frequency called a "beat frequency." This beat frequency is directly proportional to the velocity component perpendicular to the bisector of the beam intersection angle, as shown in Fig. 10-3, and in the plane of the beams.

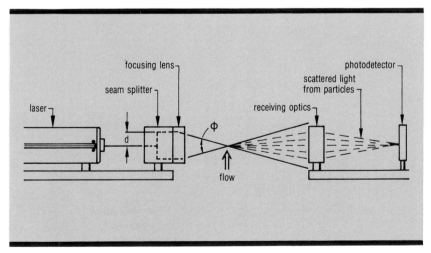

Fig. 10-2. Laser Doppler Anemometer

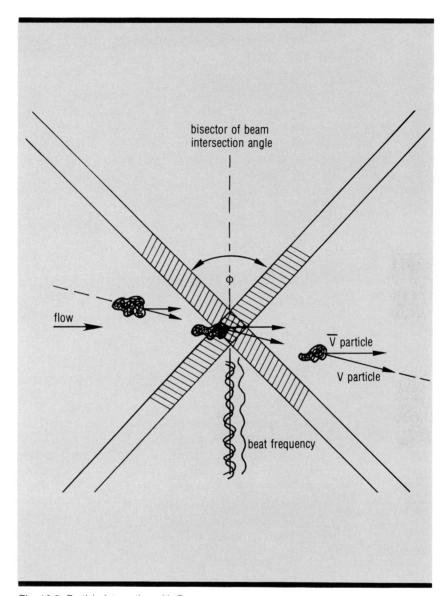

Fig. 10-3. Particle Interaction with Beams

An easier way to visualize the operation of the LDA is to think of the point at which the two beams cross as an interference zone. In this zone, light wavefronts interfere to form "fringes" or alternate regions of high- and low-intensity light. When a particle passes through the grid of variable light intensity, the scattered light as sensed by a photo detector is converted to an electrical frequency proportional to the particle velocity.

As shown in Fig. 10-4, the distance between fringes depends only on the laser-light wavelength and the angle of intersection of the light beams. Multiplying the distance between fringes by the output frequency (number of fringes passed per unit time) results in particle velocity. Consequently, no calibration is required and absolute velocity and direction are measured directly.

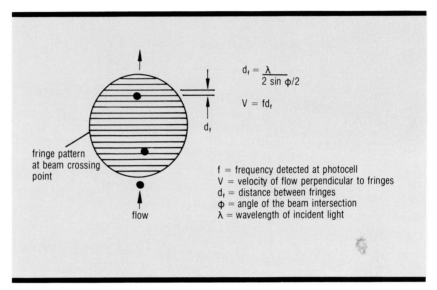

Fig. 10-4. Fringe Pattern of Variable Light Intensity

Another form of flowmeter using *existing* tags or markers is the *cross-correlation* flowmeter. In principle, cross-correlation flowmeters measure the transit time of naturally tagged signals (natural turbulence or particles, etc.) moving with the bulk velocity of the flow field. The transit time is measured between two points a known distance apart.

In order to accomplish the measurement, flow interrogation means or sensors such as ultrasound or light are used and arranged as shown in Fig. 10-5. Note, the signals from each sensor may appear to be random and poorly defined; however, utilizing the power of a *cross-correlation*, the transit time is extracted.

The cross-correlation function is defined as:

$$R_{xy}(\tau) = \frac{1}{T} \int_0^T x(t - \tau) y(t) dt$$

where τ is the cross-correlation lag time, x(t) and y(t) are the signals generated by each sensor A and B and T is the total integration time. The function $R_{xy}(\tau)$ has a maximum value when the cross-correlation lag τ is equal to the transit time $\bar{\tau}$ of the tagged signals. Consequently, the flow velocity is:

$$V = \frac{l}{\bar{\tau}}$$

where l is the spacing between sensors A and B.

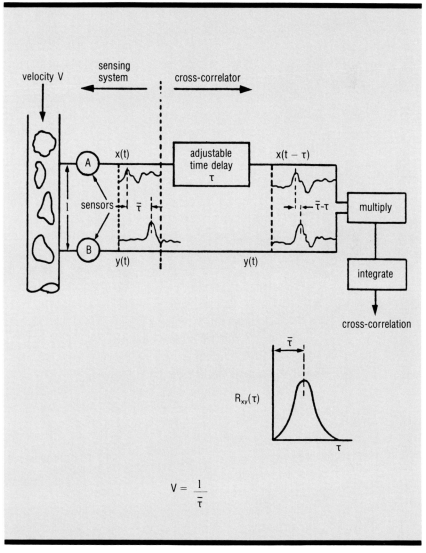

Fig. 10-5. Cross-Correlation Flowmeter

In the past, the cross-correlation calculation required elaborate and expensive computing systems. In the future and even now, the approach is becoming more realistic because of the lower cost of large-scale integrated circuits and microprocessors. Further reading materials on the different tagging techniques discussed may be found in Refs. 1 to 3.

10-2. Deflection

A technique employing both the *energy additive* and *energy extractive* approaches to flow measurement is classified here as the *deflection* technique which, in principle, utilizes some energy source which is introduced to the fluid in a pipe or duct. At zero flow, the energy source crosses the pipe or duct and the location of the received source on the other side of the pipe is noted. When flow is initiated, the flowing fluid deflects the energy source and the location of the deflected source is monitored on the opposite pipe wall. The distance between the zero flow location and the flowing location of the detected energy source is the indication of flow rate.

A unique device using the basic principles of deflection called a *fluidic deflection* flowmeter is shown in Fig. 10-6. In operation, a fluid jet, one which is compatible with the medium to be measured, is directed toward two impact-pressure ports. At zero flow in the main stream, the differential output from the impact pressure ports is zero. When the flow to be measured interacts with the jet, the jet deflects and a differential pressure is generated between the receiver ports. This differential pressure is a linear function of flow rate for a constant-density fluid. Because the device is used primarily for gas-flow measurement, density variations are of primary concern. The output signal ΔP is directly proportional to velocity, V, times the square root of the density (r) as indicated in the following equation;

$$\Delta P \sim V \sqrt{\rho}$$

The jet operates near sonic velocity and response to flow change is almost instantaneous, because the distance between the nozzle and receiver ports is only about 1/2 inch. This particular form of *deflection*-flow sensor is a point-velocity device; however, for small-line-size gas flows, average flow measurement in a pipe may be obtained.

Unit 10: Special Techniques

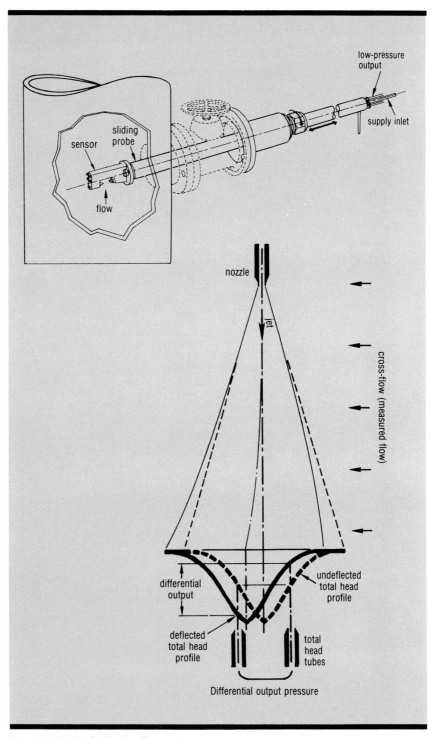

Fig. 10-6. Fluidic Deflection Flowmeter

Another form of *deflection*-technique flowmeter uses ions as the energy source. Called the *ion deflection* flowmeter, it is based on the principle that most gases may be readily ionized to form a number of positive and negative ions that differ from the other gas molecules only in their electric charges. Although the ion is influenced in the same way as the other molecules by the gas flow, it can be detected by the use of an electric field.

A gas may be ionized by means of a high-level electric field, such as an arc or corona discharge, or by the use of ionizing radiation. Some liquids also may be ionized by a radiation source. These liquids are in the nonpolar fluids class. Nonpolar fluids are liquids such as liquid hydrogen.

In one form of the corona-discharge approach to the formation of ions, a centrally located disk is used to produce ions, Fig. 10-7. Flow of gas past the disk causes an axial displacement of the radial stream of ions from the disk. The axial displacement is directly proportional to the mass flowing between the disk and the ion collector. Consequently, the representative relation between the flowmeter output and the physical properties of the flow is

$$e \sim \rho V$$

where e is an emf output, ρ is density, and V is flow rate (velocity).

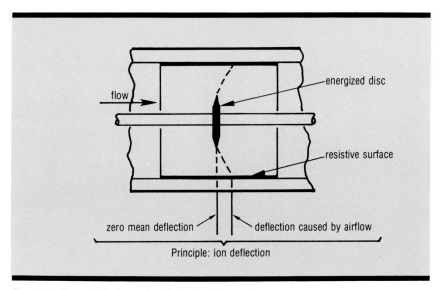

Fig. 10-7. Corona Discharge Ion Deflection Flowmeter

In another form of the *ion deflection* flowmeter, Fig. 10-8, ionization is produced by alpha radiation from a source mounted on one of the electrodes. The produced ions follow a path influenced by the gas velocity and a superimposed electric field comprised of two parts: a control or deflection field and a collection field. Motion of the ions is influenced in the deflection region by the stream velocity and the deflection field, and in the collection region by the stream velocity and the collection field. Consequently, the number of ions reaching the collector is a function of flow rate and the magnitude of the field and control voltage. By varying the control voltage, it is possible to change the range of response of the flowmeter to cover a broad range of flows with little loss in sensitivity.

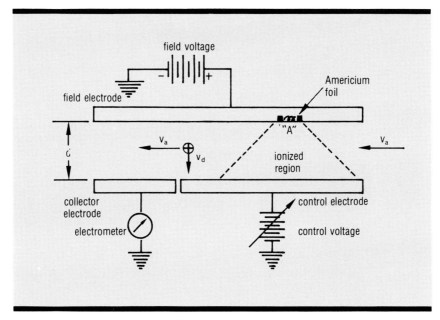

Fig. 10-8. Radiation Ion Deflection Flowmeter

Other sources of energy may also be used in the *deflection*-type flow-rate measurement technique. For example, a continuous sound wave perpendicular to a flowing fluid will deflect to some degree. For further reading on deflection devices, see Refs. 4 and 5.

10-3. By-Pass (Flow Ratio Technique)

A very convenient and economical technique to use, especially for large-line-size flow measurement, is the *by-pass* approach or *flow-ratio* technique.

As illustrated in Fig. 10-9, a small by-pass flow leg is constructed on a larger flow line. Flow is initiated in the by-pass because a pressure difference is generated across the by-pass leg by the presence of the orifice in the main line. For this system, the ratio of the flow rate in the by-pass leg to flow rate in the main line is a constant; consequently, a small flowmeter may be used in conjunction with the known constant of proportionality to establish the total flow rate in the main line.

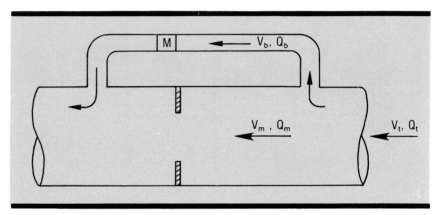

Fig. 10-9. By-Pass Flowmeter

Figure 10-9 shows an orifice plate as a restrictor; however, any differential pressure-producing device may be used in the main line, and any flow-measuring device may be used in the by-pass line. An important requirement placed on the restriction in the main line as well as the restriction inherent in the by-pass leg is that the discharge coefficient of both be constant and independent of flow rate.

This requirement will become more meaningful as the equations for the system are developed.

Consider the flow through a by-pass system such as the one illustrated in Fig. 10-9, which contains an orifice plate in the main line and a flowmeter in the by-pass line. From Eq. (5-1), Unit 5, the differential pressure developed across the orifice plate, ΔP_o, is shown to be in the following form:

$$\Delta P_o = \frac{\rho V_m^2}{2C^2} \frac{(1-\beta^4)}{\beta^4} \qquad (10\text{-}1)$$

where

ρ = density
V_m = actual velocity in main line
β = beta ratio
C = discharge coefficient

In addition to the pressure loss across the orifice, there is a natural pressure loss between the intake and exhaust ports of the by-pass due to friction on the wall of the main line acting on the flow over the distance between ports. This pressure drop may be found in the form of Eq. (4-10), Unit 4, as:

$$\Delta P_l = \frac{f\rho V^2_m}{2} \left(\frac{L}{D_m}\right) \quad (10\text{-}2)$$

Combining Eqs. (10-1) and (10-2), the total pressure loss between by-pass ports in the main line becomes;

$$\Delta P_b = \Delta P_o + \Delta P_l = \rho V^2_m \left[\frac{(1-\beta^4)}{2\,C^2\,\beta^4} + f\left(\frac{L}{D_m}\right)\right] \quad (10\text{-}3)$$

In the above equation, the term in brackets is a constant independent of flow rate if C and f are independent of flow.

Because the differential pressure ΔP_b generated between the inlet and exhaust ports of the by-pass leg initiates flow in the by-pass leg, the total pressure drop occurring in the by-pass leg must be equal to ΔP_b. *This is the important observation that allows the calculation which relates the by-pass flow rate to the flow rate in the main line.*

The pressure drop in the by-pass leg may be expressed in general form as:

$$\Delta P_b = K_b \frac{\rho}{2} V^2_b \quad (10\text{-}4)$$

where

V_b = actual velocity in by-pass line
K_b = loss coefficient for by-pass line

The loss coefficient for the by-pass line is comprised of five parts: the inlet loss, the loss coefficient of two 90° elbows, friction loss due to pipe length, the flowmeter loss coefficient, and the exit loss. It may be expressed as:

$$K_b = 2K_{90° \text{ elbow}} + K_{pipe} + K_{meter} + K_{inlet} + K_{exit}$$

Loss coefficients for elbows and other fittings as well as the friction factors, f, required to calculate the pipe loss coefficient K_{pipe} are to be found in the references of Unit 4, namely, Refs. 1 and 2.

If Eqs. (10-4) and (10-3) are combined, assuming the pressure drop ΔP_b to be the same, the final form of the equation relating the velocity in the by-pass V_b to the velocity in the main line V_m is obtained.

$$\frac{V_b}{V_m} = \left\{\frac{2}{K_b}\left[\frac{(1-\beta^4)}{2 C^2 \beta^4} + f\left(\frac{L}{D_m}\right)\right]\right\}^{\frac{1}{2}} \qquad (10\text{-}5)$$

As seen in the above equation, the velocity ratio is constant and independent of flow rate if C, f, and K_b are constant.

To obtain the total volumetric flowrate, some mathematical manipulation is required because the total flow is split between the main line and the by-pass.

$$Q_t = Q_m + Q_b$$

where

Q_t = total volume flow = $V_t A_m$
Q_m = main line volume flow = $V_m A_m$
Q_b = by-pass line volume flow = $V_b A_b$

In final form:

$$Q_b/Q_t = \frac{(D_b/D_m)^2 (V_b/V_m)}{\left[1 + (D_b/D_m)^2 (V_b/V_m)\right]}$$

and

$$V_b/V_t = \frac{V_b/V_m}{\left[1 + (D_b/D_m)^2 (V_b/V_m)\right]}$$

A good reference for pipe loss coefficients may be found in Ref. 6.

Exercises

10-1. What are two advantages of induction-tagging techniques compared with injection techniques?

10-2. Is the Laser Doppler Velocimeter an average or point-velocity instrument?

10-3. What flow marker is traced in a cross-correlation velocity-measurement technique?

10-4. Are the ion-deflection flowmeters in Figs. 10-7 and 10-8 average or point-velocity devices?

10-5. Suppose the main-line restrictor in a by-pass flowmetering system accumulates deposits causing the discharge coefficient to decrease. Is the measured total flow rate greater than or less than the actual total flow rate?

References

[1]"Fluid Meters, Their Theory and Application," *Report of ASME Research Committee on Fluid Meters*, Sixth Edition, 1971.
[2]Durst, F., Melling, A., and Whitelaw, J. H., "Principles and Practices of Laser-Doppler Anemometry," Academic Press, 1976.
[3]Beck, M. S., "Correlation in Instruments: Cross Correlation Flowmeters," *J. Phys. E: Sci. Instrum.*, Vol. 14, The Institute of Physics 1981.
[4]*FluiDynamic Devices Limited*, Bulletin 308R1079.
[5]Castle, G. S. P. and Sewell, M. R., "An Ionization Device for Air Velocity and Mass Flow Measurements," *IEEE Transactions on Industry Applications*, pg. 119, Jan./Feb. 1975.
[6]Miller, D. S., "Internal Flow Systems," Volume 5, *BHRA Fluid Engineering Series*.

Unit 11:
Mass-Flow Measurement

Unit 11

Mass-Flow Measurement

Having now been exposed to almost every conceivable type of flowmeter, you, the student, must be asking: why single out mass-flow measurement as a topic of discussion for an entire unit?

The reason is that an important distinction exists between *true* mass flow and *inferential* mass-flow measurement.

Learning Objectives—When you have completed this unit you should:

   A. Understand the differences between inferential and true mass-flow measurement.

   B. Know the basic physical principles behind each true mass-flow measurement approach.

   C. Have an appreciation for the relative merits of each approach to mass-flow measurement.

Mass-flow rate is the mass weight of a fluid flowing in a unit time. Consequently, mass-flow rate may be expressed in terms of pounds-mass per second, for example. These units of measure differ from the volume flow rate, cubic feet per second, or gallons per second, and flow rate (velocity) expressed as feet per second or meters per second in that the flow rate as well as density of the fluid determine the mass-flow rate. Hence, mass-flow rate is expressed in the following form:

$$\dot{m} = \rho A V$$

where ρ = density, pounds-mass per unit volume
   A = area, length squared
   V = flow rate (velocity), length per unit time

Because there are two measurable parameters involved, namely density and velocity, mass-flow rate may be obtained by a combination of these two measurements. This approach to mass-flow measurement is called *inferential*. For gas flow, the density is not only dependent on the gas properties, that is, type or a combination of types of gases, but is also dependent on the state of the gas, that is, pressure, temperature, and compressibility. Consequently, many measurements may be combined to form the density which only adds more credence to the *inferential* nature of the mass-flow measurement.

What then is *true* mass-flow measurement? Very simply, a *true* mass-flow measurement is one that is a direct measurement of mass and independent of the properties and the state of the fluid. How this measurement is accomplished may be found in the precepts of Newton's second law: "When an unbalanced system of forces acts on a body it produces an acceleration in the direction of the unbalanced force that varies in inverse proportion to the mass of the body."

The important concept put forth here is that mass is experienced in terms of force and acceleration. Consequently, any mechanism that combines the use of force and acceleration to measure mass may be considered a *true* mass flowmeter.

In this unit, we will cover both the *inferential* and *true* mass-flow measurement concepts most commonly found in the literature.

11-1. Inferential Mass-Flow Measurement

The most direct approach to the *inferential* measurement of mass-flow rate for any fluid is the combination of the independent measurement of density, ρ, and velocity, V, as mentioned above. Another approach not often used, however, noted in the literature and shown schematically in Fig. 11-1, is the combination of a *head*-class and *pulse*-class device to arrive at a mass-flow measurement. In principle, any density square-law dependent and linear velocity-dependent flow measurement device may be used. In practice, the head-class measurement in the form

$$\Delta P \sim \rho V^2$$

is divided by the pulse class (linear velocity-dependent) measurement in the form

$$f \sim V$$

to arrive at mass flow rate per unit area.

$$\frac{\dot{m}}{A} = \frac{\Delta P}{f} = \frac{\rho V^2}{V} = \rho V$$

Using this approach, mass-flow rate may be determined by any combination of instruments such as the orifice, venturi, target meter, etc., with any linear velocity-dependent meter such as the turbine, vortex, ultrasound, or magnetic flowmeters.

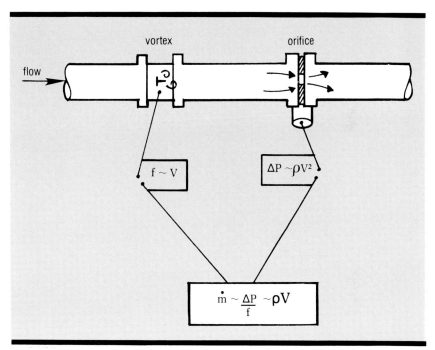

Fig. 11-1. Combined Head-Class and Pulse-Class Device for Mass-Flow Measurement

The *thermal*-class flowmeter may be considered by some to be a true mass flowmeter because the mass-flow rate is directly proportional to the heat flux. However, several other factors appear in the equation for the thermal-class flowmeter. For a quick reminder, refer to Unit 9 on powered flowmeters.

Additional terms in the equation, such as the thermal conductivity, specific heat, and viscosity which directly affect the mass-flow rate reading, cloud the picture and tend to drive the final measurement toward an *inferential* one more than a true one. For example, in cases where large changes in fluid composition and temperature are likely to occur, the measurement is definitely *inferential*. However, for gases of constant composition and small temperature variance, the mass-flow measurement is direct.

In dealing with gases, the approach most often taken is *inferential* and several measurements are taken just to obtain the gas density. For a gas, the ideal compressible-gas law provides a relationship between pressure, P, temperature, T, density, ρ, gas compressibility factor, Z, and gas composition (gas constant), R, in the form:

$$P = Z\rho RT$$

In practice, the line pressure and temperature are measured, the gas composition is measured using a gas chromatograph, for example, and the compressibility factor is obtained from tables or charts developed for the specific gas being measured. From these measurements, the density is calculated using the above equation. Once the density is known, the final calculation to obtain the mass-flow rate using a volume flow rate measurement usually taken from an orifice-plate differential-pressure measurement is a simple one.

The steps, in detail, used to calculate gas density are tedious and somewhat complex. You probably noticed throughout this ILM that detail calculations surrounding gas-flow measurement were avoided. Different approaches and philosophies taken regarding gas-flow measurement are beyond the scope of this book. For advanced reading in gas-flow measurement, the student is referred to Refs. 1 through 4.

An *inferential* approach to gas-mass-flow measurement often regarded as a standard for comparison of other measurement techniques is called the *sonic nozzle*. Other names applied to the specific devices employed are the *critical-flow nozzle* or the *choked-flow nozzle*.

In operation, a nozzle or nozzle-venturi configuration such as the one shown in Fig. 11-2 is driven to a condition where a large pressure differential exists between the throat and the nozzle inlet. For the nozzle to operate properly, the velocity at the nozzle throat must be the sonic velocity of the gas being put through the nozzle. To attain the speed of sound in the gas at the nozzle throat, usually the pressure ratio P_2/P_1 must be on the order of 1/2, the exact value being dependent on the gas composition. This critical-pressure ratio cannot be exceeded, and any further increase in the nozzle-inlet pressure will not affect the sonic velocity at the throat or the critical-pressure ratio. In short, the speed of sound in the gas is the highest velocity attainable and any further pressure changes only serve to change the mass-flow rate. Consequently, the mass flow is obtained using the following equation:

$$\dot{m} = \frac{KP_1}{\sqrt{T_1}}$$

where $\dot{m}$ = mass-flow rate
P_1 = nozzle inlet pressure
T_1 = nozzle inlet temperature
K = a calibration constant, usually supplied by the manufacturer, which includes the velocity of approach factor and the discharge coefficient.

Because critical flow is characterized by the gas velocity at the nozzle throat being equal to the speed of sound, a fixed pressure ratio exists between the inlet P_1 and throat P_2 for any inlet pressure, as long as the critical condition is maintained. Consequently, a throat pressure tap is not required and the mass-flow rate is dependent only on the upstream pressure and temperature measurements. In addition, because the velocity at the throat is sonic, downstream pressure changes, P_3, do not affect upstream pressure. However, for critical flow to be maintained in a nozzle-venturi configuration, the exit to inlet pressure ratio P_3/P_1 should be on the order of 0.8 or less.

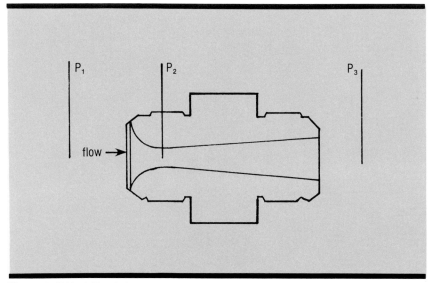

Fig. 11-2. Critical Nozzle-Venturi

11-2. True Mass-Flow Measurement

One of the important physical phenomena exploited in the measurement of true mass-flow rate is *Coriolis acceleration*. This phenomenon, found in nature, is associated with the earth's rotation and produces what is commonly referred to as a *Coriolis force*.

This force results from acceleration acting on a mass and has been experienced by anyone who dares to walk radially outward on a moving merry-go-round. As depicted in Fig. 11-3, a person walking radially outward on a rotating platform must lean toward or direct the mass of his or her moving body against the force produced by the Coriolis acceleration acting perpendicular to the radial direction of the body's motion.

Consequently, if the force acting on the body, F, the velocity of the body, V, and the angular velocity of the platform are known, the person's mass weight, M, could be calculated from the following equation:

$$M = \frac{F}{2\omega X V}$$

where X is a vector cross-product.

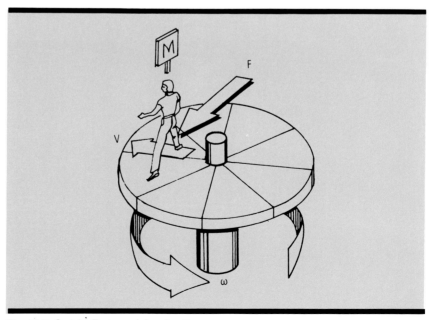

Fig. 11-3. Coriolis Force and Acceleration

In applying this phenomenon to mass-flow measurement, a mechanical structure, such as the one shown in Fig. 11-4, called the *Coriolis mass flowmeter*, is often employed. The rotor, containing metal vanes forming several radial channels, is driven at constant angular velocity ω by an external source. Any particle of fluid traveling through the radial channel with velocity V will experience the *Coriolis force*, resulting in a torque acting in the plane of rotation. The fact that the torque acts in the plane of rotation is significant, because this fact will be used to distinguish the Coriolis-type mass flowmeter from the *gyroscopic* type. With the torque acting in the plane of rotation, measurement of the torque is normally accomplished by placing a sensing means, such as a strain gage, in the drive shaft. To measure mass-flow rate, the angular velocity (motor rotational speed) is usually held constant and the torque is a direct measure of mass flow. The form of the equation is simply:

$$\dot{m} = K \frac{T}{\omega}$$

where K is a calibration constant.

In principle, this type of mass flowmeter has fast response time and may be used in situations of rapidly changing flow rates. It also is useful in the measurement of multiphase fluid flows, and some designs can accommodate reversing flows. These meters are, however, not very well suited to high flow rates, and moving parts and rotating seals may be a maintenance problem.

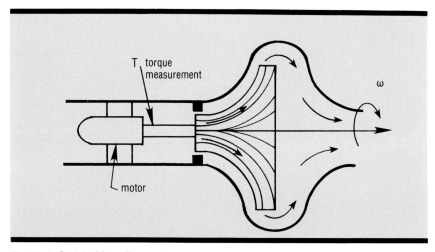

Fig. 11-4. Coriolis Mass Flowmeter

Another type of mass flowmeter using the effect of Coriolis acceleration is the *gyroscopic* mass flowmeter. Gyroscopic precession occurs in a gyroscope when a torque is applied perpendicular to the axis of rotation, Fig. 11-5. The precession is a slow rotation of the spin axis, about an imaginary line intersecting the spin axis, so as to describe a cone where the torque acts perpendicular to the cone surface.

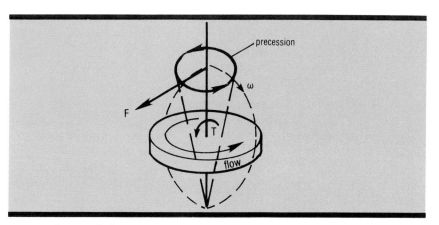

Fig. 11-5. Gyroscopic Precession

In principle, the rotating disk is similar to a fluid flowing continuously through a circular tube, Fig. 11-6. If the tube is rotated at constant angular velocity about the primary flow axis, a torque (proportional to the mass-flow rate in the tube) is produced perpendicular to the plane of rotation.

The reaction force producing the torque in a *gyroscopic* mass flowmeter is in reality a *Coriolis force*. This Coriolis force acts in the plane of rotation, ω. However, because the rotational flow path of the fluid is perpendicular to the physical plane of rotation of the tube, the torque acts in a plane perpendicular to the plane of physical rotation rather than parallel as for the *Coriolis* mass flowmeter. Consequently, a good way to distinguish a *gyroscopic* mass flowmeter from a *Coriolis* mass flowmeter is to define the plane of physical rotation relative to the plane of the acting torque. Both planes are perpendicular in the *gyroscopic* meter and parallel in the *Coriolis* meter.

The form of the final equation for the *gyroscopic* mass flowmeter is the same as for the *Coriolis* meter:

$$\dot{m} = K \frac{T}{\omega}$$

where, of course, K is a different calibration constant.

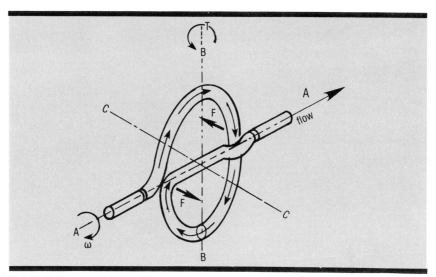

Fig. 11-6. Gyroscopic Mass Flowmeter

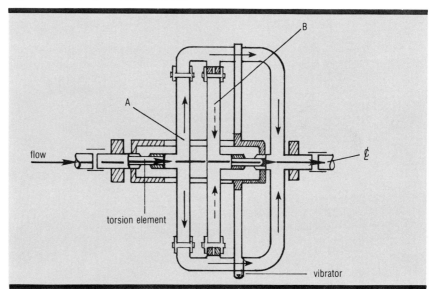

Fig. 11-7. Oscillating Coriolis Mass Flowmeter

A very advantageous means of eliminating the need for rotating seals and rotating parts is to generate a *Coriolis force* through vibration or oscillation of the flow tube. Such a device is described in detail in Ref. 5 and shown in the schematic, Fig. 11-7. Two identical tubes, A and B, are constructed so that both may be oscillated at constant frequency and amplitude about the axis of symmetry or primary flow axis. Tube A is the measuring tube through which all the fluid passes and tube B is the mass flow compensation mechanism containing the stationary fluid. Fluid flowing in the oscillating measuring tube

A exerts a variable force on the tube walls, thereby creating an oscillating torque in the plane of oscillation. The torque produced is comprised of three components: the torque resulting from the *Coriolis force*, which is directly proportional to mass flow, the torque component arising due to the inertia of the tube, and the torque component arising from the inertia of the fluid in the tube which is proportional to the fluid density. For the *Coriolis flowmeter* operating with constant angular velocity, the inertia effects do not exist. However, in a oscillating system, the angular velocity is varying in magnitude and sign, thereby causing the resulting torque measurement to be dependent not only on mass flow but also the moment of inertia of the system. To compensate for the torque produced by the inertial components of the system, a similar tube, B, filled with stationary fluid is used such that the moment of inertia of tube B matches that of tube A, and the resulting output is subtracted from the total output of tube A leaving then only the torque sensitive to mass flow.

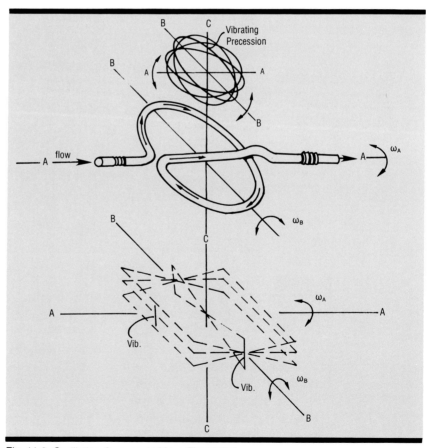

Fig. 11-8. Oscillating Gyroscopic Mass Flowmeter

The *gyroscopic* principle may also be exploited in the oscillatory mode. In geometric form, the *oscillating gyroscopic flowmeter* is similar to the one employing constant angular rotation. As shown in Fig. 11-8, the circular tube, having two degrees of freedom, one about axis A-A and the other about axis B-B is put into a vibratory mode usually at the resonant frequency of the system. Once in oscillation, the circular tube, due to the gyroscopic effect, goes into a vibration precession, or a motion similar to a nutating disk, about the axis C-C.

The amplitude of these precessional oscillations not only is proportional to the mass flow of the fluid but also is dependent on the moment of inertia of the circular tube and the inertia of the fluid.

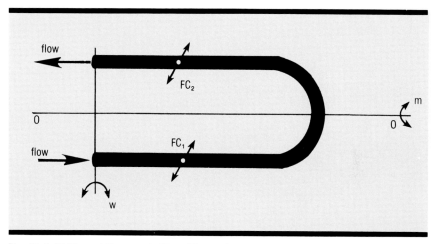

Fig. 11-9. U-Shaped Gyroscopic Mass Flowmeter

Another form of device employing the *gyroscopic* principle of operation in the oscillatory mode is shown in Fig. 11-9. A tube shaped in the form of the letter U forms one-half of the classical *gyroscopic mass flowmeter*. An electromagnetic oscillator drives the U-shaped, tuning-fork-like structure at the resonant frequency of the system, thereby producing a Coriolis acceleration and resultant force. The force acts alternately (perpendicular to the flow path) in opposite directions, as shown in Fig. 11-9, causing an oscillating moment about axis O-O of the flowmeter. The resulting moment m, acting about the central axis and in a plane perpendicular to the driving moment w, produces a twist-type motion, shown in Fig. 11-10, where the deflection angle θ is directly proportional to the mass-flow rate for a constant angular velocity.

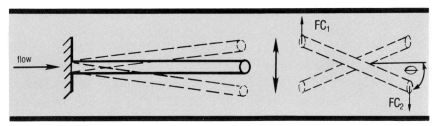

Fig. 11-10. Motion of U-Shaped Flowmeter

Because the angular velocity is not constant, inertial effects have to be considered. The inertial effects are essentially eliminated by the detection scheme which employs an optical sensor to measure the deflection angle near the center position of the U-shaped pipe excursion. The center position is the location where the angular acceleration of the pipe is near zero, that is, the point of constant angular velocity, and any inertial effects are eliminated in the flow signal.

In laboratory tests, a device of this design, described in greater detail in Ref. 6, produced mass-flow measurements with a total accuracy of ±0.2% of full-scale flow when employed with fluid having a specific gravity in the range from 0.5 to 2.5, and flow rates rom 0.1 to 25 pounds mass per minute. In addition, the flowmeter output is linear to within ±0.2% of full scale over the entire range of the testing.

The *angular-momentum* type of mass flowmeter has long been one of the more popular approaches to *true* mass-flow measurement. A sketch showing the basic form of the device is presented in Fig. 11-11. In operation, an upstream impeller is driven at constant angular velocity by a motor source. As fluid enters the impeller, it takes on the rotational velocity of the impeller and upon exiting has an angular velocity equal to that of the impeller. The fluid is then confronted with a turbine, which may rotate about the same axis as the impeller, but is constrained from rotation by a calibrated spring or some form of restraining torque. Upon leaving the turbine, all the angular momentum in the fluid has been transferred to the turbine and the torque produced in the turbine is directly proportional to mass flow.

The equation describing the mass-flow relation to torque and angular velocity is of the form:

$$\dot{m} = \frac{T}{K\omega}$$

where K is a constant and ω, the impeller angular velocity is held constant.

Another means of obtaining the mass-flow rate by using the same device is to keep the torque T constant by changing the angular velocity of the impeller using feedback control. Then the impeller angular velocity, or the impeller motor revolutions per minute become the measure of mass flow.

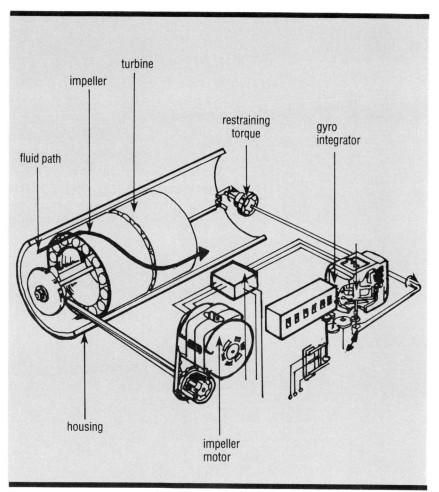

Fig. 11-11. Angular-Momentum Mass Flowmeter

Because the *angular-momentum* concept in mass-flow measurement is, in fact, a transfer of energy imparted by the mass of rotating fluid from the driving impeller to the driven turbine, in theory it is possible to measure the mass flow by sensing the power used in driving the impeller, thereby eliminating the need for a turbine.

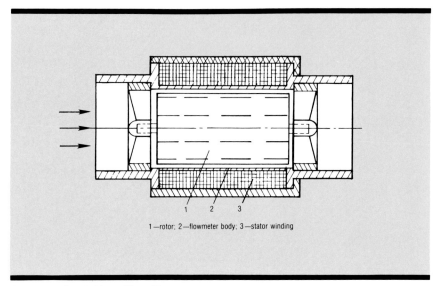

Fig. 11-12. Single-Rotor Mass Flowmeter

This type of mass flowmeter, Fig. 11-12, has one impeller constructed in the form of a rotor for an electric motor, for example, and a stator winding is placed on the outside of the flow tube. A relationship between power, P, and mass flow exists such that:

$$P = \dot{m}\,\omega^2 K$$

where ω is the angular velocity of the rotor and K is a constant. Consequently, the power transmitted to the fluid stream by the constant angular rotation of the impeller is directly proportional to mass flow.

Mass flowmeters of the *angular-momentum* type generally have the same potential accuracy as the *Coriolis-gyroscopic* type. A disadvantage of the angular-momentum type over the oscillatory Coriolis-gyroscopic type is the rotating parts which may prove to be high-maintenance items.

A novel form of *true* mass-flow measurement system comprised solely of differential producing or *head*-class devices is called the *pressure-differential* type. Two schemes, one employing venturi tubes and the other orifice plates, are shown in the schematic drawing of Fig. 11-13. Both approaches are exactly the same theory and the final form of the mass-flow equation is the same for both.

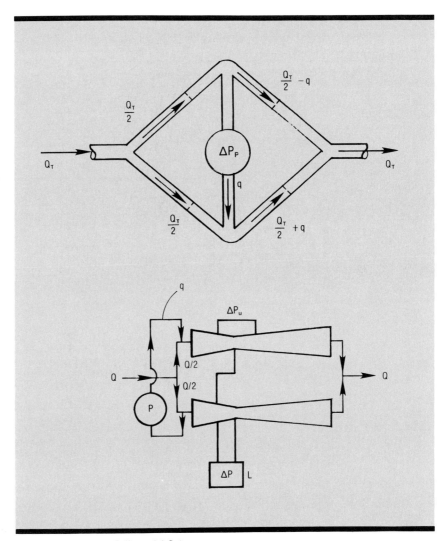

Fig. 11-13. Pressure-Differential Schemes

The technique employed is based on pumping a portion of the flow from one part of a fluid-flow system into another part of the same system at a constant and known flow rate. As seen in Fig. 11-13, a simple system employing four identical orifice plates in a balanced bridge arrangement may be set up where the differential pressure across the upper an lower legs of the system has the following form:

$$\Delta P_U \sim C\rho Q^2_U$$
$$\Delta P_L \sim C\rho Q^2_L$$

If the pump is operated such that a constant volume flow, q, is transferred from the upper leg to the lower leg, the following equations apply:

$$Q_U = \frac{Q_T}{2} - q$$

$$Q_L = \frac{Q_T}{2} + q$$

Now considering that the pressure rise across the pump ΔP_P must be equal to the pressure difference in both legs, we obtain:

$$\Delta P_P = \Delta P_L - \Delta P_U \sim C\rho \, (Q_T/2 + q)^2 - C\rho(Q_T/2 - q)^2$$

or $\Delta P_P \sim 2C\rho 2 Q_T q$

Since mass flow $\dot{m} = \rho Q_T$, then:

$$\dot{m} \sim \frac{\Delta P_P}{2Cq}$$

Consequently, if the pumped flow is held constant, and the discharge coefficients of the orifice installation are the same, the mass flow is directly proportional to the differential pressure measured across the pump.

Similarly, if the flow is divided equally between two identical venturi tubes, as shown in Fig. 11-13, and a pump extracts flow from one tube and injects that quantity of flow into the other tube at a constant rate, the difference in differential pressure measured for each venturi is proportional to mass flow.

$$\Delta P_U \sim C\rho(\frac{Q}{2} + q)^2$$

$$\Delta P_L \sim C\rho(\frac{Q}{2} - q)^2$$

$$\Delta P_L - \Delta P_U = 2C\rho Q q$$

$$\dot{m} = \frac{\Delta P_L - \Delta P_U}{2Cq}$$

Again, certain assumptions are inherent in the above analysis. It is assumed the upstream pressure is not disturbed by the fluid transfer, that is, the extraction and injection processes and the loss coefficients for both legs in the system are identical. Both requirements are difficult to achieve in practice and any deviation from constancy results in a direct error in the mass-flow measurement.

Exercises

11-1. True mass-flow measuring techniques are:
(a) energy additive
(b) energy extractive
(c) both
(d) neither

11-2. Is the sonic nozzle an inferential or true mass-flow measuring device?

11-3. What is the disadvantage of angular-momentum-type mass flowmeters? Oscillatory mass flowmeters?

11-4. What type of relationship exists between the mass-flow rate and the indicating parameter in all true mass flowmeters described?

11-5. A flow through an orifice ($\beta = .6$, $a = .8$, see Unit 4) produces 5 psi of differential pressure. The same flow through a vortex flowmeter ($K = 3 \frac{hz}{fps}$) causes a shedding frequency of 20 hz.

(a) If the line size is 3", what is the mass flow rate?
(b) Is this an inferential or true measurement?

References

[1]*Orifice Metering of Natural Gas*, Gas Measurement Committee Report No. 3, American Gas Association, Inc., 1969.
[2]*Fuel Gas Energy Metering*, Gas Measurement Committee Report No. 5, American Gas Association, Inc., 1970.
[3]Miller, R. W., *Flow Measurement Engineering Handbook*, McGraw-Hill Book Co. (1983)
[4]Ward-Smith, A. J., *Internal Fluid Flow*, Clarendon Press, Oxford, 1980.
[5]Katys, G. P., *Continuous Measurement of Unsteady Flow*, The Macmillan Company, 1964.
[6]Plache, K. O., "Coriolis/Gyroscopic Flowmeter," *Mechanical Engineering*, March 1979.

Unit 12:
Flowmeter Selection

Unit 12

Flowmeter Selection

This unit will summarize the entire course in the form of a thought process for the selection of the best flowmeter for a specific application.

To accomplish this task, a series of charts is presented outlining the step-by-step thought process; however, no definitive selection chart is provided. The reason that no single, all-inclusive chart is provided is simple: *flowmeter selection* is too complex to fit in a two-dimensional chart. *Flowmeter selection* is a complex thought process sometimes requiring several iterations to arrive at the *best selection*.

Learning Objectives—When you have completed this unit, you should:

    A. Understand the selection process and how to use it to determine the best approach to flow metering in a particular situation.

    B. Be able to select the best flow-metering device for use in a particular situation.

12-1. Flowmeter Selection Process

A suggested thought process leading to flowmeter-selection criteria is presented in the chart in Fig. 12-1. In order to go through the selection chart in the direction indicated by arrows, a series of questions must be asked and answers obtained before one may proceed to the next step. For example, to enter the chart, one must ask the first question.

> **Am I interested in a *measurement* of the fluid in my process or am I interested in the *control* of that fluid flow?**

This question is the most important question in that it has a severe impact on the final selection criterion, which we will define as *smiles per dollar*. A discussion on the value of this criterion is given later in this unit.

Let us first assume that our interest lies in *measurement*. The next question to ask is: **am I interested in *mass*, *volume*, or *rate* flow?** Most often, if the basic interest is in a measurement, either mass or volume flow is desired. For example, if I am buying thousands of gallons of fuel oil my interest is in the total gallons delivered, not in the rate at which it is delivered, i.e., the velocity of the flow. If the fluid I am purchasing varies in density during the delivery process, then my interest is in the total mass of fluid delivered, i.e., pounds of fluid.

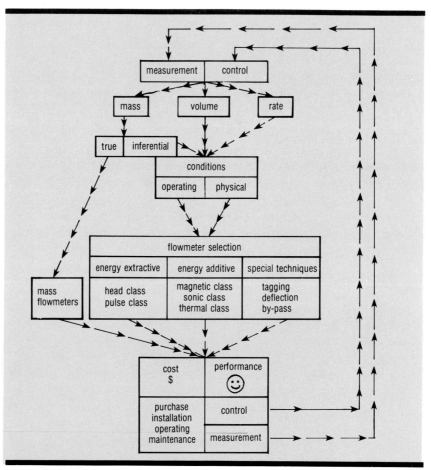

Fig. 12-1. Thought Process for Flowmeter Selection

On the other hand, if my interest is in process *control*, I may only need a flow *rate* measurement, although one may need to control flow using volume or mass flow. Think of the final criterion in this case. If an insertion probe, such as the pitot tube, could be employed to deliver a velocity flow rate at a single point in a duct, the cost would be far less and overall performance may be equivalent for control purposes to, say for example, a venturi tube which provides the volumetric flow rate. Consequently, the required *performance* (smiles) divided by the *cost* (dollars) is far greater if the pitot tube is employed.

Therefore, at the outset, two very important decisions must be made in the *thought process* for *flowmeter selection—measurement* or *control* and *mass, volume,* or *rate*.

Referring again to Fig. 12-1, let us assume a *mass* flow measurement is desired. In this case, the question to be asked is: **do I need a *true* mass-flow measurement or will an *inferential* measurement satisfy my needs?** If a true mass-flow measurement is required, the selection process is relatively simple because there are only a limited number of devices to choose from, see Unit 11. Again, the criterion for the selection is *smiles per dollar*.

If an *inferential* mass-flow measurement is sufficient, the next step to take is the consideration of the flow measurement *conditions*. A statement of the *conditions* (i.e., the *operating* and *physical* conditions the meter is to experience) is a necessary and important step in the selection process for volume and rate, as well as mass flow.

In Fig. 12-2, a chart containing some of the operating and physical conditions to be considered is provided. There may be many more conditions to be considered depending on the particular process.

Some of the basic questions to ask of the *operating conditions* are:

    Is the fluid a *liquid* or a *gas*?

    What are the maximum and minimum flow rates?

    Is the flow in the *laminar* or *turbulent* flow regime? (Calculate Reynolds number.)

    What are the properties of the fluid?

    Is it corrosive? dirty? a slurry? non-Newtonian? two-phase?

    What is the temperature? viscosity? pressure?

Some of the basic questions to ask of the *physical conditions* are:

What is the line size?

What does the installation look like?

What are the upstream pipe conditions?

Is the flow in an open channel?

Is the pipe only partially filled?

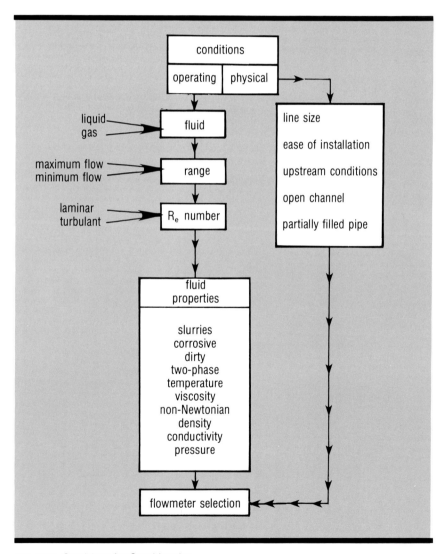

Fig. 12-2. Conditions for Consideration

The combination of answers to questions regarding both the physical and operating conditions provides the information base to be used in the selection process. The actual selection process involves everything you have learned about flow measurement in Units 2 through 11; however, to simplify the discussion on the thought process for flowmeter selection, only performance criteria in terms of percentage are presented in Fig. 12-3.

Note that the flowmeter selection chart, Fig. 12-3, is categorized in terms of the approach to flow measurement in that the devices are listed as either *energy extractive, energy additive,* or *special techniques.* Each device in each class of flowmeter has associated with it an overall system accuracy, A, in terms of percent of full scale, S, or in terms of actual flow rate, R, and a repeatability or statement of precision, P, in terms of percent of actual flow rate.

For a review of the performance terminology, i.e., (accuracy—uncertainty) and (repeatability—precision), refer to Unit 3 on *General Flow-Measurement Terminology.*

The actual values listed in the chart of Fig. 12-3 are intended to be nominal values for comparing one device against another. Actual values may vary depending on the manufacturer and/or the primary standard of measurement used to evaluate the device.

Both the system accuracy, A, and the repeatability, P, are listed in the selection chart so that a clear distinction is made between the value of a device as a *measurement* tool or a *control* tool. For example, a single pitot probe may only provide a total volume flow measurement accuracy of ±2% of the full scale flow, thereby representing a poor approach to *measurement*. However, a precision on the order of ±1/2% of flow rate may be adequate for your control purposes and provide you with a high *smile-per-dollar* ratio.

At this point, once a device has been selected either on the basis of *measurement* or *control* or both, the evaluation process can begin. In the evaluation process, both *performance* and *cost* must be considered. As shown in Fig. 12-4, the priorities on performance criteria are different for a device used strictly for measurement as opposed to a device used only for control.

Flowmeter Selection

Energy Extractive

Head Class	A	P	Pulse Class	A	P
Conventional			**Positive Displacement**		
Square-edge orifice	±¾S	±¼	Liquid sealed drum	±½R	±⅛
Eccentric orifice	±1½S	±¼	Two diaphragm	±½R	±⅛
Segmental orifice	±2S	±½	Reciprocating piston	±½R	±⅛
Quadrant-edge orifice	±2S	±¼	Rotary piston	±½R	±⅛
Venturi	±¾S	±¼	Nutating disk	±½R	±⅛
Nozzle	±¾S	±¼	Rotary vane	±½R	±⅛
			Lobed impeller	±½R	±⅛
Special			**Current**		
Target	±¾S	±¼	Propeller	±1S	±¼
Elbow	±2S	±½	Cup anemometer	±1S	±¼
Pitot	±2S	±½	Vane anemometer	±1S	±¼
Variable area	±2S	±½	Turbine	±½R	±⅛
Linear resistance	±3S	±½	Gas turbine	±¾R	±⅛
Open Channel			**Fluid Dynamic**		
Weirs	±3R	±¾	Vortex	±¾R	±¼
Flumes	±5R	±¾	Precessing vortex	±2R	±¼
			Fluidic oscillator	±1S	±¼

Energy Additive

Magnetic Class	A	P
Alternating current	±½R	±¼
Pulsed direct current	±½R	±¼
Sonic Class		
Time-of-flight	±1R	±¼
Doppler	±1S	±¼
Thermal Class		
Thermo-anemometers	±2R	±½
Calorimetric	±2S	±½

Special Techniques

Tagging	A	P
Flow markers	±1R	±½
Laser Doppler	±½R	±¼
Correlation	±2R	±½
Deflection		
Fluidic	±2S	±½
Ion	±2R	±¼
By-Pass		
Nominal values	±1S	±½

A = system accuracy/uncertainty
S = percent of full scale accuracy
R = percent of actual flow rate accuracy
P = precision/repeatability

Fig. 12-3. Flowmeter Selection Chart

For example a *measurement* device, especially one used in billing (custody transfer), would have overall accuracy as a high-priority item, whereas a *control* device may not require high overall-system accuracy. A control device having good linearity is attractive. Linearity is a high-priority item in a control device because it implies good precision over wide range. Linearity in a measurement device may not be too important if the range of measurement is small.

In general, (accuracy—uncertainty), resolution, and (repeatability—precision) are three high-priority performance criteria for *measurement* devices, whereas, (repeatability—precision), resolution, and linearity are high-priority criteria for *control* devices.

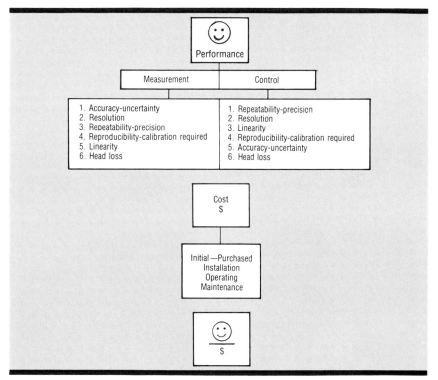

Fig. 12-4. Performance and Cost Considerations

Head loss is shown as a low-priority item for both measurement and control devices, however this evaluation factor has significant impact on operating cost. Cost considerations as listed in Fig. 12-4 are (initial—purchased) costs, and installation, operating, and maintenance costs. With regard to flowmeter cost, a graph of relative cost as a function of line size is provided in Fig. 12-5. This graph is intended only as a guide to the relative total cost of devices in the different classes and techniques or approaches to flow measurement. Consequently, the only real distinction on a cost basis is clear in the large-line-size category. In small-line sizes below eight-inches pipe diameter, there are several overlaps in cost and the cost criteria become less clear. The only way to obtain a valid initial cost for a device is from the manufacturer. To that initial cost, installation, operating, and maintenance costs must be added.

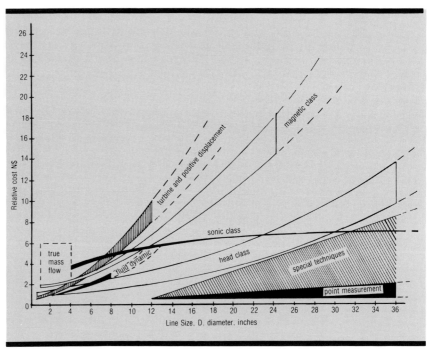

Fig. 12-5. Relative Cost of Flowmeters

Once you have collected your thoughts on the relative merits of the flowmeter selection, considering the *operating* and *physical* conditions, and you have come to grips with *performance* relative to your application of *measurement* and/or *control*, and you have considered the cost of ownership, you are now ready to form an opinion, in your own mind, as to the value of one flow-measurement device over another on the basis of the *smiles-per-dollar* ratio. The dollars may be calculated with a high degree of confidence. The smiles are somewhat subjective and to a great extent dependent on relative terms. It is for this reason that there may be more than one flowmeter device chosen as the best for a particular application.

In conclusion, as the chart shown in Fig. 12-1 indicates, the flowmeter selection is normally an iteration process and may have to be done several times before converging on the *best* flowmeter or flowmeters to consider in a particular application.

12-2. Example of Flowmeter Selection

In order to demonstrate the flowmeter-selection thought process, an example will be given to dramatize some of the important considerations and value judgments required.

Example

The Dry Gulch Mining Company, located in a remote desert area, buys water from the Desert Water Corporation for their mining operations. Because of the high cost of water, both companies want the most accurate flow-measurement device for use in billing.

The first step in the thought process is easy in this case in that a *measurement* device is essential.

Proceeding to the second step and asking the question, "Am I interested in *mass, volume,* or *rate?*," requires some thought. Because past history indicates the delivered water temperature is relatively constant (so that the density variance is no greater than ±0.05%) no consideration is given to a *true-mass* flow measurement. However, water temperature will continue to be monitored so that periodic inferential mass-flow measurements may be made using the measured temperature, water density tables, and the *volumetric* flow rate. A point velocity for flow *rate* measurement is not acceptable because of questionable velocity profile changes over the four-to-one range of water delivery.

In the third step, the *conditions* to which the flowmeter will be subjected, both *physical* and *operating,* are listed.

Operating

a) Fluid—liquid, water
b) Maximum flow — 4,000,000 gal/day
   Minimum flow — 1,000,000 gal/day
   Range 4 : 1

c) Pipe Reynolds number—turbulent region
   Maximum — 976,580
   Minimum — 244,145
d) Fluid properties
   Temperature — 60°F
   Pressure—25 psig
   Density—62.37 lbm/ft³
   Viscosity — 0.755×10^{-3} lbm/sec-ft
   Conductive—Yes
   Dirty—No

Physical

a) Line size — 8 inch
b) Installation—easy
c) Upstream conditions—good
d) Filled pipe—yes

At this point, flowmeter selections are made and the evaluation process begins with the *performance* and *cost* analysis.

For an 8-inch flowmeter, a rough estimate of the initial purchase cost, Fig. 12-5, and best accuracy, Fig. 12-3, for five type of flowmeters gives the following:

	Cost N$	Accuracy
1. Turbine and positive displacement	4.5 to 5.5	±1/2%R
2. Magnetic class	3.8 to 4.4	±1/2%R
3. Sonic class	5	±1%R
4. Fluid dynamic	2.3 to 3.2	±3/4%R
5. Head class	1.5 to 2.5	±3/4%S

Upon inspection, comparing the best expected accuracy, ±1/2% of rate, we find the turbine, positive-displacement, and magnetic-class devices are equivalent, however the cost could be significantly greater for turbine or positive-displacement flowmeters. Because the fluid is water (and conductive), a magnetic-class flowmeter may be used.

Comparing the two remaining highest estimated cost meters — the magnetic class and the sonic class — we find we can get twice the performance in terms of accuracy from the magnetic class.

Thinking of other cost factors, such as installation, operating, and maintenance, several thoughts should enter your mind. For example, turbine and positive-displacement meters may have higher operating costs due to higher pressure loss and subsequent higher pumping costs than the obstructionless magnetic or sonic flowmeters. In addition, maintenance costs may be higher because of inherent moving parts.

Although installation costs may be less for the sonic class, especially if a clamp-on device is considered, the expected inaccuracies due to velocity profile changes may be greater than the nominal ±1% of rate listed.

At this point, we should feel comfortable listing three devices to look at more closely and proceed with the final stage of flowmeter selection.

Based on a first iteration of the *smiles-per-dollar* ratio, the three flowmeters selected for further evaluation are the *magnetic flowmeter*, the *vortex flowmeter*, and the *venturi flowmeter*.

A venturi is chosen over the orifice or nozzle strictly on the basis of head loss. The lower head loss in the venturi means operating cost savings in terms of less expended energy in pumping.

Consider the following *performance-cost* analysis for the three candidate flowmeters. At the current rate of $0.50 for 1,000 gallons of water, we calculate the potential misunderstanding between buyer and seller in terms of dollars per year, assuming a 300-day operating year.

	Magnetic	Vortex	Venturi
Maximum flow	±$3000	±$4500	±$4500
Minimum flow	±$ 750	±$1125	±$4500

In the worst case, the buyer may demand a $4,500 return from the seller and the seller may feel he should receive an additional $4,500 from the buyer. Under the best conditions of minimum flow consumption and highest accuracy, the discrepancy is only ±$750 for the year. The dollar difference at minimum flow between the magnetic flowmeter and the venturi demonstrates the attractiveness of a flowmeter with percent of actual accuracy specifications.

Now let's compare the potential yearly savings if the highest accuracy device is chosen.

Potential yearly savings	Magnetic	Vortex	Venturi
Maximum	0	$1,500	$1,500
Minimum	0	$ 375	$3,750

At minimum flow consumption in using the venturi for one year, there is enough money in question, $3,750, to pay for a large part of a magnetic flowmeter and more than enough for a vortex flowmeter. However, if the magnetic flowmeter is chosen over the vortex flowmeter, at maximum flow there may be over four years to the break-even point, or at minimum flow consumption, 15 years.

Here, the choice between magnetic and vortex flowmeters begins to require a more detailed analysis and consideration of other *performance* and *cost* factors.

Let us list the performance factors in order of priority and the actual cost of ownership.

A review of magnetic and vortex flowmeter manufacturers' literature provides the following information on *performance*.

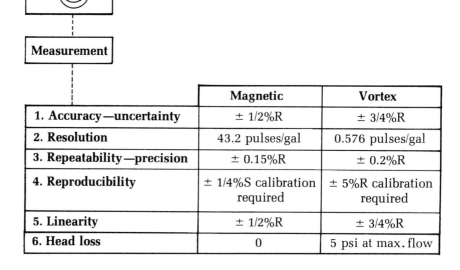

	Magnetic	Vortex
1. Accuracy—uncertainty	± 1/2%R	± 3/4%R
2. Resolution	43.2 pulses/gal	0.576 pulses/gal
3. Repeatability—precision	± 0.15%R	± 0.2%R
4. Reproducibility	± 1/4%S calibration required	± 5%R calibration required
5. Linearity	± 1/2%R	± 3/4%R
6. Head loss	0	5 psi at max. flow

Based on total *accuracy* alone, the magnetic flowmeter is slightly better, however as we just calculated, the questionable cost savings are minimal.

The *resolution* of the magnetic flowmeter is better and may be a significant performance factor. For example, if it is necessary to increase or decrease the *volumetric* flow rate by say 50 gallons per minute, the vortex meter delivers 288 pulses per minute (using a ten times multiplyer) whereas the magnetic flowmeter delivers 2,160 pulses per minute. Consequently, in any one minute integration time, the vortex meter will have ±3.5% error and the magnetic flowmeter will have a ±0.05% error, because pulses can be discriminated to only ±1 count.

Repeatability or *precision* of both meters is excellent, and at a constant flow rate over a long time, both meters should deliver excellent results.

Reproducibility is a measure of the characteristic nature of any flowmeter to be taken off the shelf, installed in a flow line and deliver a specified performance level. In this case, both meters have unacceptable levels of meter factor *reproducibility* and both require calibration before delivery.

The *linearity* of both meters is excellent and follows the overall accuracy specification.

A very important performance factor, in that it affects the cost of ownership, is the *head loss*. The magnetic flowmeter being obstructionless has minimal head loss, whereas the vortex meter has an approximate 5 psi head loss at maximum flow conditions.

Keeping these *performance* factors in mind, let us now list the *cost* factors and try to resolve in our minds the final *smiles-per-dollar ratio*.

A call to a local flowmeter manufacturer and a little effort in the way of simple cost calculations result in the following information on *cost*.

	Magnetic	Vortex
Initial—purchased	$6,300	$3,300
Installation	$ 480	$ 480
Operating	$ 13 per year	$ 392 per year
Maintenance	$1,970 per year	$1,970 per year

Maintenance cost is calculated on the basis of two hours per month at a cost of $30 per hour. In addition, a one-day-per-year downtime is assumed at an average delivery rate so that one full day's delivery charge is in question. The maintenance cost is the same for both flowmeters.

Operating cost is calculated on the following basis: for the vortex flowmeter, the yearly pumping cost is calculated from the *head loss*, average flow rate, and cost of electricity, based on a 300-day year:

$$\text{Yearly pumping cost} = \left[3.82 \, \frac{Q \times \Delta P \times C}{E} \right] \frac{300}{365}$$

where

$Q = 1736$ gal/min (average flow rate)
$\Delta P = 1.8$ psi at average flow rate
$C = \$0.03$ per Kw-hr
$E = 0.75$ pump-efficiency factor

For the magnetic flowmeter, the yearly cost of electricity to power the flowmeter is calculated, based on 60 watts power consumption:

$$\text{Yearly electricity cost} = 8.76 \, (W \times C) \frac{300}{365}$$

where $W = 60$ watts
$C = \$0.03$ pr Kw-hr

For *installation* costs, let us assume it takes 16 man-hours at $30 per hour.

In the final *cost* analysis, the one-time costs and the continuing costs are compared.

	Magnetic	Vortex
One-time cost	$6,300	$3,300
Continuing cost	$1,983 per year	$2,362 per year

The relative *cost of ownership* of the vortex flowmeter over the magnetic flowmeter is $379 per year; however there is a $3,000 initial difference in the one-time cost.

Based on this information, which flowmeter has the better smiles-per-dollar ratio?

$$\frac{\smiley}{\$} = ?$$

Exercises

Select a flowmeter for each of the following situations:

12-1. The nominal flow rate of a gravel/water slurry in a 14-inch line must be measured. Accuracy is low priority.

12-2. A food processing plant requires an accurate volume-measuring device to repeatedly deliver a fixed quantity of fluid to each container. The fluid is highly viscous, nonconductive, and non-Newtonian and must not be contaminated.

12-3. The flow rate of a dirty sediment-laden liquid in a 30-inch line must be controlled. Accuracy is low priority.

12-4. The flow rate of a sludgy, nonconducting fluid in a three-inch line must be accurately controlled.

Appendix A: Suggested Readings and Study Materials

Appendix A

Suggested Readings and Study Materials

Independent Learning Modules

Bibbero, R. E., *Microprocessors in Industrial Control* (Instrument Society of America, 1983).
Cho, Chun H., *Measurement and Control of Liquid Level* (Instrument Society of America, 1982).
Driskell, Les, *Control-Valve Selection and Sizing* (Instrument Society of America, 1982.)
Murrill, P. W., *Fundamentals of Process Control Theory* (Instrument Society of America, 1981).
Shinskey, F. G., *Controlling Multivariable Processes* (Instrument Society of America, 1981).

Handbooks

"Flow of Fluids through Valves, Fittings, and Pipes," Technical Paper No. 410, pp. A-23 to A-25 (Crane Co., 4100 S. Kedzie Ave., Chicago, IL 60632).
Miller, R. W., *Flow Measurement Engineering Handbook* (McGraw-Hill Book Co., 1983).

Textbooks

Katys, G. P., *Continuous Measurement of Unsteady Flow* (A Permagon Press Book, Macmillan Co., 1964).
Fluid Meters, Their Theory and Application, Report of the ASME Research Committee on Fluid Meters, Sixth Ed., 1971.
Durst, F., Melling, A., and Whitelaw, J. H., *Principles and Practices of Laser-Doppler Anemometry* (Academic Press, 1976).
Miller, D. S., *Internal Flow Systems*, Vol. 5, BHRA Fluid Engineering Series.
Ward-Smith, A. J., *Internal Fluid Flow* (Clarendon Press, Oxford, 1980).

Technical Magazines and Journals

"A History of Flow Measurement by Pressure-Difference Devices," Publication 1010/460 (George Kent Limited).
L. F. Moody, "Friction Factors for Pipe Flow," *Transactions of the ASME*, Vol. 66, Nov. 1944, pp. 671-678.
"The Venturi Water Meter," by Herschel Clemens, *Transactions of the American Society of Civil Engineers*, 1887.
"Measurement of Natural Gas," by Thos. R. Weymouth, *Transactions of the ASME*, 1912, p. 1091.
"Discharge of Elastic Fluids under Pressure," Wm. Froude, Proceedings, Inst. of Civil Engineers (Great Britain), 1847, Vol. 6.
Curran, D. W., "Laboratory Determination of Flow Coefficient Values for the Target-Type Flowmeter at Low Reynolds Number Flow," *FLOW, Its Measurement and Control in Sciences and Industry*, Vol. 2 (Instrument Society of America, 1981).
Kiel, G., "Total Head Meter with Small Sensitivity to Yaw," NACA-TM 775, Aug., 1935.
Barker, M., "On the Use of Very Small Pitot Tubes for Measuring Wind Velocity," Proc. of Royal Soc., Series A., Vol. 101, p. 435, 1922.
Kindsvater, C. F., and Carter, R. W., "Discharge Characteristics of Rectangular Thin Plate Weirs," *Transactions of the ASCE*, Vol. 124, p. 772, 1959.

Smith, E. S., "The V-Notch Weir for Hot Water," *Transactions of the ASME*, Vol. 56, p. 787, 1934; Vol. 57, p. 249, 1935.

Parshall, R. L., "The Parshall Measuring Flume," Colorado Agricultural Experimental Station Bulletin 423, Mar. 1936.

Palmer, H. K., and Bowlus, F. D., "Adaptation of Venturi Flumes to Flow Measurement in Conduits," *Transactions of the ASCE*, Vol. 101, p. 1195, 1936.

Cousins, T., Nicholl, A. J., "Comparison of Turbine and Vortex Flowmeters," CME, Feb. 1978.

Dijstelbergen, H. H., "The Performance of a Swirl Flowmeter," *J. Physics*, 3(11)886-888, 1970.

Miller, R. S., DeCarlo, J. P., and Cullen, J. T., "A Vortex Flowmeter—Calibration Results and Application Experiences, NBS Special Publication 484, Vol. 2, pp. 549-570.

Inkley, F. A., Walden, D. C. and Scott, D. J., "Flow Characteristics of Vortex Shedding Flowmeters, Measurement and Control," Vol. 13, May 1980.

Adams, R. B., *A Fluidic Flowmeter*, ISA Reprint 73-815.

Kolin, A. "An Alternating Field Induction Flowmeter of High Sensitivity," *Review of Scientific Instruments*, Vol. 16, p. 109, May 1945.

Fath, J. P. "Magmeters with Ultra-Stable Zero," ISA 75-830, 1975.

Powell, D. J. "Ultrasonic Flowmeters Basic Designs, Operation, and Application Criteria," *Plant Engineering*, Vol. 33, No. 9, p. 93, May 1979.

Schmidt, T. R. "What You Should Know about Clamp-On Ultrasonic Flowmeters," *InTech* 0192-303X/81/05/059/04/, May 1981.

Waller, J. M. "Guidelines for Applying Doppler Acoustic Flowmeters," *InTech* 0192-303X/80/10/0055/03, Oct. 1980.

Freymuth, P. "Review: A Bibliography of Thermal Anemometry," *Transactions of the ASME Journal of Fluids Engineering*, Vol. 102, p. 152, June 1980.

Beck, M. S., "Correlation in Instruments: Cross-Correlation Flowmeters," *J. Phys. E: Sci. Instrum.*, Vol. 14, (The Institute of Physics, 1981).

Castle, G.S.P., and Sewell, M. R., "An Ionization Device for Air Velocity and Mass Flow Measurements," *IEEE Transactions on Industry Applications*, p. 119, Jan./Feb. 1975.

"Orifice Metering of Natural Gas," Gas Measurement Committee Report No. 3 (American Gas Association, Inc., 1969).

"Fuel Gas Energy Metering," Gas Measurement Committee Report No. 5 (American Gas Association, Inc., 1970).

Plache, K. O., "Coriolis/Gyroscopic Flowmeter," *Mechanical Engineering*, Mar. 1979.

Standards

Glossary of Terms Used in the Measurement of Fluid Flow in Pipes, ANSI/ASME MFC-1M-1979.

Measurement of Fluid Flow by Means of Orifice Plates, Nozzles, and Venturi Tubes Inserted in Circular Cross-Section Conduits Running Full, ISO 5167-1980 (E).

Appendix B: Solutions to All Exercises

Appendix B

Solutions to All Exercises

UNIT 2

2-1. The following meters use the extractive energy approach:

Weir	Cup anenometer	Elbow meter
Target	Oval gear	Turbine
Orifice plate	Flume	Nozzle
Pitot probe	Fluidic oscillator	Linear resistance
Area meter	Propeller	Venturi
Vortex	Nutating disk	Liquid seal meter

2-2.
Head Class		**Pulse Class**
Weir	Elbow meter	Liquid seal meter
Target	Nozzle	Vortex
Orifice plate	Linear resistance	Cup anenometer
Pitot probe	Venturi	Oval gear
Area meter		Fluidic oscillator
Flume		Propeller
		Turbine
		Nutating disk

All of the pulse-class meters have a frequency output.

2-3.
Magnetic (AC or DC)	Magnetic class
Time-of-flight	Sonic class
Doppler	Sonic class
Thermo-anenometer	Thermal class
Calorimetric	Thermal class

2-4. The orifice plate, Venturi, and nozzle are conventional head-class meters.

UNIT 3

3-1. The *secondary device* on a head-class meter is usually a differential-pressure-measurement device (a D/P cell). There are exceptions. On a target meter, for example, the secondary device is a force or motion sensor which measures the drag force or displacement of the target. The meter, however, still works on the principle of differential pressure (or head). The force on the target is the product of the differential pressure (pressure in front minus pressure behind) and the target area.

3.2. (a) The pulse class meter recorded the following volume flow:

$$\text{Volume} = (907 \text{ pulses}) \div (53.68 \text{ pulses/cu ft})$$
$$= 16.89 \text{ cu ft}$$

(b) The weight of water collected is the difference between the final (*gross*) weight and the original (*tare*) weight:

$$\text{Weight} = \text{gross weight} - \text{tare weight}$$
$$= 3017.5 \text{ lb} - 2008.2 \text{ lb}$$
$$= 1009.3 \text{ lb}$$

The volume of water is its weight divided by its density:
$$\text{Volume} = \text{weight/density}$$
$$= (1009.3 \text{ lb})/(62.4 \text{ lb/cu ft})$$
$$= 16.17 \text{ cu ft}$$

(c) Actual volume = 16.17 cu ft
Observed volume = 16.89 cu ft

$$\% \text{ error} = \frac{16.89 - 16.17}{16.17} \times 100 = +4.45\%$$

(d) Actual volume flow rate = Actual vol/elapsed time
= 16.17 cu ft/211 sec
= 0.0766 cu ft/sec
Observed volume flow rate = Observed vol/elapsed time
= 16.89 cu ft/211 sec
= .080 cu ft/sec

$$\text{Error} = \frac{.0800 - .0766}{.0766} \times 100 = 4.44\%$$

This error should be exactly the same as that in part (c). The difference between these errors is due to calculation (round-off) errors: if all calculations were carried out to 10 decimal places, the results would be closer.

(e) The mass-flow rate is the volumetric flow rate multiplied by the density:
$$\dot{m} = \rho Q$$

(f) This is a *static weighing* proof test.

3-3. (a) Reynolds number $= \dfrac{\rho V D}{\mu}$

$$= \frac{(1.20 \text{ kg/m}^3)(1 \text{ m/s})(.06\text{m})}{(1.8 \times 10^{-5} \text{ kg/m·s})}$$

$$= \frac{(.072 \text{ kg/ms})}{1.8 \times 10^{-5} \text{ kg/ms}} = 4000$$

(b) This Reynolds number indicates transitional or partially turbulent flow (remember, below 2000 indicates *laminar*, and above 10,000 indicates *turbulent*).

(c) The ratio must be constant as long as the units of the quantities are dimensionally consistent (the ratio is truly dimensionless). This must be checked:

$$\text{Re} = \frac{\rho VD}{\mu} = \frac{(\text{lbm/ft}^3)\,(\text{ft/s})\,(\text{ft})}{(\text{lbm/ft sec})} = \frac{\text{lbm/ft sec}}{\text{lbm/ft sec}} = dimensionless$$

Thus, this expression will also yield the (dimensionless) Reynolds number. If the diameter was expressed in inches, then it would have to be converted to feet before the Reynolds number could be computed.

3-4. (a) If 95% of the measurements are in the ±0.5% error band, then 5% are outside of it. Statistically half should be high and half low; so you should expect 2.5% of the measurements to be more than 0.5% high.

(b) These specifications assume that the error is both *random* and *systematic* in nature.

3-5. (a and b) Meter I can be in error by up to 0.5% of its full scale value. (The word *span* in this specification is misleading: the span of a meter is not equal to the full scale value unless the meter reads down to zero flow). Thus it can be in error by (.005) (100 ft/sec) = 0.5 ft/sec. Meter II can be in error by up to 0.5% of the actual measurement. In Meter III the output is related to the input by a multiplicative factor (flow = meter factor × raw reading). The meter is calibrated and the average meter factor is found. Both the *systematic error* and *random error* cause the meter factor to vary in this case up to ±0.5%. This specification is the same as that for Meter II, except that in this case it specifies the error relative to a mean rather than the actual "true" value.

Now, on with the answers:

Meter	Reading	Maximum error	Minimum actual flow rate	Maximum actual flow rate
Meter I	20 ft/sec	0.5 ft/sec	19.5 ft/sec	20.5 ft/sec
	60	0.5	59.5	60.5
	100	0.5	99.5	100.5
Meter II	20	0.1	19.9	20.1
	60	0.3	59.7	60.3
	100	0.5	99.5	100.5
Meter III	20	0.1	19.9	20.1
	60	0.3	59.7	60.3
	100	0.5	99.5	100.5

(c) The specifications for Meters I and II are common ways of indicating primarily *random* errors (it is assumed in this specification that a proof test will be run to experimentally determine the meter factor, hence eliminating any systematic errors resulting from an inaccurate meter factor). The specification of Meter III is used to specify *random* and *systematic* errors.

(d) The range and rangeability are the same for all three meters. The *range* is the difference between the maximum and minimum values that the meter can measure, and the *rangeability* is the ratio of these numbers:

Range = 100 ft/sec − 20 ft/sec = 80 ft/sec
Rangeability = 100 ft/sec ÷ 20 ft/sec = 5:1

UNIT 4

4-1. (a) The orifice diameter is
$$d = \beta D = (0.60)(4.00 \text{ in}) = 2.40 \text{ in}$$

(b) The velocity of approach factor is
$$E = 1/\sqrt{1-\beta^4} = 1/\sqrt{1-(0.60)^4} \sim 1.072$$

(c) The theoretical velocity is found from Eq. (4-4):
$$V_t = \beta^2 E \sqrt{2\Delta P/\rho} = (.60)^2(1.072)\sqrt{\frac{2(5.00 \text{ lb}_f/\text{in}^2 \times 32.2 \text{ lb}_m/\text{slug} \times 144 \text{ in}^2/\text{ft}^2}{62.4 \text{ lb}_m/\text{ft}^3} \frac{\text{lb}_m}{\text{lb}_f}}$$
$$V_t = 10.52 \text{ ft/sec}$$

(d) The actual velocity is
$$V_a = CV_t = 0.615(10.52 \text{ ft/sec}) = 6.47 \text{ ft/sec}$$

(e) The diameter of the pipe cross section is D = 4 inches = .333 feet, so its area is $A_1 = \pi(.333)^2/4 = .0871 \text{ ft}^2$. The density of the water is $\rho = 62.4 \text{ lb}_m/\text{ft}^3$.

Therefore,
$$Q_a = V_a A_1 = (6.47 \text{ ft/sec})(0.0871 \text{ ft}^2) = 0.564 \text{ ft}^3/\text{sec}$$
$$\dot{m}_a = V_a A_1 \rho = (6.47 \text{ ft/sec})(0.0871 \text{ ft}^2)(62.4 \text{ lb}_m/\text{ft}^3) = 35.2 \text{ lb}_m/\text{sec}$$

4-2. (b)
$$V_a = \varepsilon \beta^2 CE \sqrt{\frac{2 \cdot \Delta P}{\rho}}$$

$$V_a = (0.935)(0.6)^2(0.615)\left(\frac{1}{\sqrt{1-0.6^4}}\right)\sqrt{\frac{2(5 \times 144 \text{ psF})(32.2 \frac{\text{lb}_m}{\text{slug}})}{\left(62.4 \frac{\text{lb}_m}{\text{ft}^3}\right)}}$$

$$V_a = 174.5 \text{ Fps}$$

again,

$$A_1 = 0.0871 \text{ ft}^2$$

$$Q = A_1 V_a = (0.0871 \text{ ft}^2)(174.5 \text{ Fps}) = 15.2 \text{ cFs}$$

$$\dot{m} = \rho Q = (0.075 \frac{\text{lb}_m}{\text{ft}^3})(15.2 \text{ cFs}) = 1.14 \frac{\text{lb}_m}{\text{sec}}$$

Appendix B: Solutions to All Exercises   249

4-3. (a)
$$Re_D = \frac{VD\rho}{\mu} = \frac{(10.52 \text{ Fps})(0.333 \text{ ft})(62.4 \frac{lb_m}{ft^3})}{\left(6.7 \times 10^{-4} \frac{lb_m}{ft-\text{sec}}\right)}$$

$Re_D = 326{,}264 \rightarrow$ turbulent flow

(b) Using Darcy's Formula:
$$\Delta P = \frac{\rho f L V^2}{2 D g_c} = \frac{(62.4 \frac{lb_m}{ft^3})(0.027)(2000 \text{ ft})(10.52 \text{ Fps})^2}{2(0.333 \text{ ft})(32.2 \frac{lb_m}{\text{slug}})}$$

$\Delta P = 17{,}389 \text{ psF} = 121 \text{ psi}$

(c) $Re_D = \dfrac{(0.2 \text{ Fps})(\frac{1}{12} \text{ ft})(62.4 \frac{lb_m}{ft^3})}{(6.7 \times 10^{-4} \frac{lb_m}{ft-\text{sec}})} = 1552$

$\rightarrow$ laminar flow

Using the laminar flow equation:
$$\Delta P = \frac{32 \mu L V}{D^2 g_c} = \frac{32(6.7 \times 10^{-4} \frac{lb_m}{ft-\text{sec}})(2000 \text{ ft})(0.2 \text{ Fps})}{(0.083 \text{ ft})^2 (32.2 \frac{lb_m}{\text{slug}})}$$

$\Delta P = 38.7 \text{ psF} = 0.27 \text{ psi}$

5-1. (a) $62.11 \frac{lb_m}{ft^3}$

(b) $0.0005115 \frac{lb_m}{ft-\text{sec}}$

(c) $V_{max} = \dfrac{\dot{m}_{max}}{\rho A} = \dfrac{(50 \frac{lb_m}{ft-\text{sec}})}{(62.11 \frac{lb_m}{ft^3}) \frac{\pi}{4} (\frac{4.026}{12} \text{ ft})^2} = 9.1 \text{ Fps}$

(d) $\Delta P_{max} = \rho \dfrac{g}{g_c} \Delta h_{max} = (62.11 \frac{lb_m}{ft^3}) \left(\dfrac{32.2 \frac{ft}{\text{sec}^2}}{32.2 \frac{lb_m}{\text{slug}}}\right) \dfrac{600}{12} \text{ ft} = 3105.5 \text{ psF}$

(e) $Re_D = \dfrac{\rho V D}{\mu} = \dfrac{(62.11 \frac{lb_m}{ft^3})(9.1 \text{ Fps})(\frac{4.026}{12} \text{ ft})}{0.0005115 \frac{lb_m}{ft-\text{sec}}} = 368000$

5-2.

(a) Using Eq. 5-1: $\quad C = \dfrac{V\sqrt{1-\beta^4}}{\beta^2 \sqrt{\dfrac{2 \cdot \Delta P}{\rho}}}$

$$\dfrac{\beta^4}{1-\beta^4} = \dfrac{V^2}{C^2\left(\dfrac{2\,\Delta p}{\rho}\right)} = \dfrac{(9.1\text{ Fps})^2}{(0.6)^2 \left[\dfrac{2(3105.5\text{ psF})}{\left(\dfrac{62.11\text{ lb}_m/\text{ft}^3}{62.2\text{ lb}_m/\text{slug}}\right)} \right]}$$

$\beta^4 = \dfrac{0.072}{1 + 0.072} \quad \to \beta = 0.51$

(b) $\text{Re}_d = \dfrac{\text{Re}_D}{\beta} = \dfrac{368{,}000}{0.51} = 722{,}000$

(c) Table 5-2(b) gives $C = 0.6040$

(d) $C = 0.5959 + 0.0312(0.51)^{2.1} - 0.184(0.51)^8$

$\quad + 0.0029(0.51)^{2.5} \left(\dfrac{10^6}{722000}\right)^{0.75}$

$\quad + 0.039\left(\dfrac{1}{4}\right)(0.51)^4(1-0.51^4)^{-1} - 0.0337\left(\dfrac{1}{4}\right)(0.51)^3$

$C = 0.5959 + 0.0076 - 0.0008 + 0.0007$

$\quad + 0.0007 - 0.0011 = 0.6714$

(e) $\Delta P_{max} = \dfrac{(9.1\text{ fps})^2(1-0.51^4)(62.11\,\dfrac{\text{lb}_m}{\text{ft}^3})}{2(0.51)^4(0.6714)^2(32.2\,\dfrac{\text{lb}_m}{\text{slug}})}$

$\Delta P_{max} = 2444\text{ psF} = 17\text{ psi}$

5-3.

(a) $C = 0.984$

(b) As in Exercise 5-2 (a).

$\dfrac{\beta^4}{1-\beta^4} = \dfrac{(9.1\text{ fps})^2}{(0.984)^2 \left[\dfrac{2(3105.5\text{ psF})}{\left(\dfrac{62.11\text{ lb}_m\text{ft}^3}{62.2\text{ lb}_m/\text{slug}}\right)} \right]} = 0.0266$

Appendix B: Solutions to All Exercises

$$\beta^4 = \frac{0.0266}{1 + 0.0266} \rightarrow \beta = 0.40$$

(c) $Re_d = \dfrac{Re_D}{\beta} = \dfrac{368000}{0.40} = 920000$

The acceptable ranges are

$2 \times 10^5 < Re_d < 2 \times 10^6$ and,

$0.3 < \beta < 0.7$

(d) $\Delta P_{max} = \rho \dfrac{g}{g_c} \Delta h_{max} = (62.11 \dfrac{lb_m}{ft^3}) \left(\dfrac{32.2 \dfrac{ft}{sec^2}}{32.2 \dfrac{lb_m}{slug}} \right) \left(\dfrac{120}{12} ft \right)$

$\Delta P_{max} = 621.1 \text{ psF} = 4.31 \text{ psi}$

$$\dfrac{\beta^4}{1 - \beta^4} = \dfrac{(9.1 \text{ Fps})^2}{(0.984)^2 \left[\dfrac{2(621.1 \text{ psF})}{\left(\dfrac{62.11 \text{ lb}_m/ft^3}{62.2 \text{ lb}_{m/slug}} \right)} \right]} = 0.133$$

$$\beta^4 = \dfrac{0.133}{1 + 0.133} \rightarrow \ = \beta = 0.585$$

(e) $\beta = 0.59$

$$\Delta P_{max} = \dfrac{\rho V^2_{max} (1 - \beta^4)}{2\beta^4 C^2} = \dfrac{\left(\dfrac{62.11 \dfrac{lb_m}{ft^3}}{32.2 \dfrac{lb_m}{slug}} \right) (9.1 \text{ fps})^2 (1 - 0.59^4)}{2(0.59)^2 (0.984)^2}$$

$\Delta P_{max} = \ 598.7 \text{ psF } = 4.16 \text{ psi}$

5-4.

(a) Using Eq. 5-1

$$\dfrac{\beta^4}{1 - \beta^4} = \dfrac{(9.1 \text{ fps})^2 (62.11 \text{ lb}_m/ft^3)}{(0.96)^2 (2) (621.1 \text{ psF}) (32.2^{\text{ lb}_m}/slug)} = 0.140$$

$$\beta^4 = \dfrac{0.140}{1 + 0.140} \rightarrow \beta = 0.59$$

(b) $C = 0.9965 - 0.00653(0.59)^{1/2} \left[\dfrac{10^6}{368{,}000} \right]^{1/2} = 0.9882$

(c) $\dfrac{\beta^4}{1-\beta^4} = \dfrac{(9.1 \text{ fps})^2 (62.11 \, \frac{lb_m}{ft^3})}{(0.9882)^2 (2) (621.1 \text{ psF}) (32.2 \, \frac{lb_m}{slug})} = 0.1317$

$\beta^4 = \dfrac{0.1317}{1 + 0.1317} \rightarrow \beta = 0.584$

(d) $Re_d = \dfrac{Re_D}{\beta} = \dfrac{368{,}000}{0.584} = 630{,}000$

Both are in the acceptable range.

(e) $\beta = 0.59$

$C = 0.9965 - 0.00653 (0.59)^{1/2} \left[\dfrac{10^6}{368.000} \right]^{1/2} = 0.9882$

$\Delta P_{max} = \dfrac{\rho V_{max}^2 (1-\beta^4)}{2\beta^4 C^2} = \dfrac{(62.11 \, \frac{lb_m}{ft^3}) (9.1 \text{ fps})^2 (1 - 0.59^4)}{(32.2 \, \frac{lb_m}{slug}) (2) (0.59)^4 (0.9882)^2}$

$\Delta P_{max} = 593.6 \text{ psF} = 4.12 \text{ psi}$

UNIT 6

6-1. (a) The formula for an area meter is

$\dot{m} = (a^2 - 1)C\sqrt{2gAF\rho} = K_1\sqrt{A}$ where K_1 constant

Since h is proportional to A, let $a = K_2 h$, or

$\dot{m} = K_1 \sqrt{K_2 h}$

Combining the constants (by introducing a new constant)

$K = K_1 \sqrt{K_2}$ gives us

$\dot{m} = K\sqrt{h}$

Appendix B: Solutions to All Exercises   253

(b) The force F is exerted on the fluid by a piston of area $\pi D^2/4$. Thus, the fluid must exert a differential pressure

$$\Delta P = F/(\pi D^2/4) = 4F/\pi D^2$$

to counteract the force of the piston.

6-2. (a) The problem will be solved using Eq. (6-2). First, calculate the velocity of the flow:

$$Q = AV \text{ or } V = Q/A$$

$$V = (2\text{cfs})/\left[\frac{6}{12}\right]^2 \frac{\pi}{4} = 10.2 \text{ fps}$$

The elbow flowmeter equation requires that the fluid density be in units of slugs/ft³. To convert lb_m/ft^3 to slug/ft³, divide by $g_c = 32.174$.

$$\rho = 62.4 \text{ lb}_m/\text{ft}^3/32.174 = 1.94 \text{ slugs/ft}^3$$

$$V = \sqrt{\frac{R}{D}\frac{\Delta P}{\rho}} \text{ or } \Delta P = \frac{V^2 D \rho}{R}$$

$$\Delta P = \frac{(10.2)^2 \left(\frac{6}{12}\right)(1.94)}{\left(\frac{6}{12}\right)} = 201.8 \text{ psf} \quad (1.4 \text{ psi})$$

(b) Doubling the radius results in one-half the differential pressure. Note the inverse relationship between ΔP and R in the equation.

$$\Delta P = \frac{(10.2)^2 \left(\frac{6}{12}\right)(1.94)}{\left(\frac{12}{12}\right)} = 100.9 \text{ psf} \quad (0.7 \text{ psi})$$

6-3. The value of C is not known without first knowing the Reynolds number. Since the velocity is unknown, one must iterate.

STEPS

(1) Select a median value for C: Let C = 0.5

(2) Calculate $\dot{m}$:

$$\dot{m} = (0.5)\left[\frac{\left(\frac{.957}{12}\text{ ft}\right)^2 - \left(\frac{.7656}{12}\text{ ft}\right)^2}{\frac{.7656}{12}\text{ ft}}\right]\sqrt{\frac{\pi}{2}(32.174\frac{lb_m}{\text{slug}})(.40 \text{ lb}_f)(62.4\frac{lb_m}{\text{ft}^3})}$$

$$\dot{m} = (.5)(.0359)(35.5) = 0.637 \text{ lb}_m/\text{sec}$$

(3) Calculate V: $\dot{m} = \rho V A \rightarrow V = \dfrac{\dot{m}}{\rho A} = \dfrac{0.637 \dfrac{\text{lb}_m}{\text{sec}}}{(62.4 \dfrac{\text{lb}_m}{\text{ft}^3})\dfrac{\pi}{4}\left(\dfrac{.957}{12}\text{ft}\right)^2} = 2.04$ fps

(4) Calculate Re_D: $Re_D = \dfrac{VD}{\nu} \dfrac{(2.04)(.957/12)}{1.41 \times 10^{-5}} = 1.15 \times 10^4$

(5) Using Fig. 6-4, find C corresponding to the calculated Re_D: C = 0.70

(6) Recalculate in: m = (.70)(.0359)(35.5) = 0.892 lb_m/sec
Now, since C = 0.70 for $Re_D > 3000$, no further iterations are necessary. (Note that if the actual velocity and Reynolds number are desired, they must be recalculated from the new mass-flow rate.)

6-4. Using Eq. (6-9), the theoretical velocity is

$$V_{actual} = V_{theoretical} \times C = \dfrac{CD^2 \times \Delta P}{32 \, \mu \, L}$$

$$\Delta P = \Delta h \rho \dfrac{g}{g_c} = (0.024_m)(13,600 \dfrac{\text{Kg}}{\text{m}^3})(9.8 \dfrac{\text{m}}{\text{sec}^2})$$

$$\Delta P = 3144 \text{ N}/_m 2$$

$$V_a = \dfrac{(0.90)(0.002\text{m})^2 (3144\text{N}/_m 2)}{32 (1.31 \times 10^{-3} \dfrac{\text{N-sec}}{\text{m}^2})(0.30\text{m})} = 0.90 \text{ m/s}$$

6-5. The deflection of the manometer is a measure of the dynamic head of the water relative to the boat hull. Equate these and solve for V:

$$\tfrac{1}{2} \rho_{H_2O} V^2 = \rho_{H_g} g \Delta h, \text{ and since } \rho_{H_g} = 13.55 \, \rho_{H_2O}$$

$$\tfrac{1}{2} \rho_{H_2O} V^2 = 13.55 \, \rho_{H_2O} (32.2 \text{ ft/s}^2)(\dfrac{1.5}{12} \text{ ft }), \text{ solving for V:}$$

$$V = \sqrt{\dfrac{2(13.55)(32.2)(1.5)}{12}} = 10.4 \text{ fps}$$

Appendix B: Solutions to All Exercises   255

UNIT 7

7-1. Using Figure 7-4:

(a) 30° triangular weir.

(b) The maximum head should not exceed one-third the crest width; more conservatively, $0.1 \leq h \leq 1.7$ ft limits the flow to 30 cfs.

7-2. Solution: $q_a = 3.33 (L - 0.2 H_u) H^{3/2}$

$L = 2$ ft $H_u = .5$ ft

$q_a = 3.33 (2 - 0.2(.5))(.5)^{3/2} = 2.24$ cfs

7-3. The discharge equation for a triangular weir is derived in the same manner as that for the rectangular weir.

The differential discharge is $dq = L\, V\, dh$

and $V = \sqrt{2gh}$

But now, L varies with h

$L = 2(H-h)\tan\dfrac{\theta}{2}$

where: H = total head

h = distance down from H

and θ = notch angle

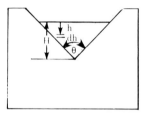

$$q_t = \int_0^H dq = \int_0^H 2(H-h)\tan\left(\dfrac{\theta}{2}\right)\sqrt{2gh}\, dh = 2\sqrt{2g}\,\tan\left(\dfrac{\theta}{2}\right)\int_0^H (H-h)\sqrt{h}\, dh$$

$$q_t = 2\sqrt{2g}\,\tan\dfrac{\theta}{2}\left[\int_0^H H\, h^{1/2} dh - \int_0^H h^{3/2} dh\right]$$

$$q_t = 9.26\left[\dfrac{2}{3} H\, h^{3/2} - \dfrac{2}{5} h^{5/2}\right]_0^H = 9.26\left[\left(\dfrac{2}{3} - \dfrac{2}{5}\right) H^{5/2}\right]$$

$q_t = 2.47\ H^{5/2}$

7-4. (a) $\frac{H_b}{H_u} \times 100 = \%$ submergence, so $H_b = \frac{(1.5)(66.7)}{100} = 1.0$ ft

(b) Since the submergence is less than 70%, the flow is free and Fig. 7-7 can be used.

An eight-foot flume with an upper head of 1.5 ft will pass 58 cfs.

(c) Head loss $= H_u - H_b = 1.5 - 1.0 = 0.5$ ft.

7-5. (a) Enter Fig. 7-7 with $H_u = 4.25$ ft and $Q = 1000$ cfs. A 25-foot flume satisfies these conditions. As a check, use the results from Fig. 7-7 in the equations on page 122.

$Q = (3.6875(25) + 2.5)(4.25)^{1.6} = 958 \approx 1000$ cfs. This checks within the readable accuracy of Fig. 7-7.

(b) Head loss $= H_u - H_b$

$H_u = 4.25$ ft

$H_b = (4.25)(.40) = 1.7$ ft

Head loss $= 4.25 - 1.70$ ft $= 2.55$ ft

UNIT 8

8-1.

	Positive displacement	Current type	Fluid-dynamic
(a) Requires no moving parts			✓
(b) Sensitive to approach conditions		✓	✓
(c) High % slip		✓	
(d) Frequency of pulses linearly related to velocity	✓	✓	✓
(e) Counts discrete fluid parcels	✓		
(f) Nutating disk	✓		
(g) Turbine flowmeter		✓	
(h) Cup anemometer		✓	

8-2. The orifice/dp system responds with $\Delta P \sim V^2$. Therefore, the differential pressure output will span over a 400 to 1 range.

The turbine flowmeter responds linearly to velocity changes so the pulse-frequency output will span a 20 to 1 range.

8-3. (a) $S = \dfrac{fh}{V} = 0.127$, $V = 1$ fps

$h = (0.3)(4/12 \text{ ft}) = 0.1$ ft

$f_1 = 1.27 \dfrac{\text{pulses}}{\text{sec}}$

$Q = VA = (1\text{fps}) \dfrac{\pi}{4} (\dfrac{4}{12} \text{ ft})^2 = 0.087$ cf/sec

Meter factor, $K = \dfrac{f_1}{Q} = \dfrac{1.27 \frac{\text{pulses}}{\text{sec}}}{0.087 \frac{\text{cf}}{\text{sec}}} = 14.6 \dfrac{\text{pulses}}{\text{cf}}$

(b) Each pulse represents 1/K cubic feet of flow,

or $\dfrac{1}{14.6 \frac{\text{pulses}}{\text{cf}}} = 0.0685 \dfrac{\text{cf}}{\text{pulse}}$

This is called the *resolution* of the flowmeter.

(c) $f_{20} = \dfrac{(0.127)(20 \text{ fps})}{0.1 \text{ ft}} = 25.4$ pulses/sec

8-4. The correct choice is a positive-displacement type flowmeter. Fluid dynamic flowmeters require Reynolds numbers well into the turbulent regime to deliver a high degree of performance, that is, on the order of ±1% of reading and are not well suited for highly viscous flow measurement. Current-type flowmeters infer flow rate from the reaction of the fluid on the propeller and, therefore, the response is nonlinear for non-Newtonian fluids. Positive-displacement flowmeter performance is not affected by either of these conditions.

8-5. A fluid dynamic flowmeter is the best choice. The presence of sand in the flow is likely to jam or cause increasing slip in a positive-displacement flowmeter. Current-type meters require finely machined and lubricated bearing surfaces which are adversely affected by errosive contaminants.

UNIT 9

9-1. The constant magnetic field produces a constant polarity at the electrodes. This results in an accumulation of gases on the electrodes which alters the resistance of the electric circuit thereby introducing error, or destroying the measurement.

9-2. Pulsed-DC magnetic flowmeters operate with stable zero references unaffected by variations present in electrical supply lines. In addition, most of the extraneous information (*noise*) can be eliminated by subtracting the signal received between pulses from that received during a pulse, greatly improving flow-signal quality.

9-3. The Doppler techniques measure the difference in frequency between a single transmitted sonic wave and its reflections received from particles moving with the fluid.

The "time-of-flight" technique requires that two sonic signals be sent into the flow for a single measurement; one with, the other against the flow direction. This technique also requires that the fluid be free of particles to avoid diffusion of the sonic beam.

9-4. A high frequency is desirable since better resolution can be obtained. Frequencies on the order of 1 Megahertz are typical of this technique.

9-5. Hot-film probes are easily adapted to shapes suited for special applications. They are more rugged and respond better over a wider frequency range than hot-wire probes.

9-6. Thermo-anemometers infer flow rate by monitoring the heat lost from the wire or film immersed in the fluid. The control volume is the wire or film. Calorimetric flowmeters monitor the heat gained by the fluid passing through the meter. The control volume is the fluid.

UNIT 10

10-1. Induction tagging techniques do not contaminate the fluid and, in most cases, the process line need not be penetrated.

10-2. The LDV measures point velocity.

10-3. Cross-correlation techniques trace natural flow markers such as turbulence or particles in the fluid.

10-4. The ion-deflection flowmeters described measure average velocity.

10-5. A decrease in the discharge coefficient increases the differential pressure existing between the by-pass inlet and outlet. This results in an increase of the actual by-pass ratio and therefore a higher reading on the flowmeter. The system will then indicate a total flow rate greater than the actual total flow rate.

UNIT 11

11-1. (c) Both. Mechanical energy is added and removed in true measurement techniques.

11-2. The sonic nozzle is an inferential device. The mass flow depends on both the pressure and the temperature at the inlet. Undetected variations in the metered medium also will introduce error, since the sonic speed depends on the gas composition.

11-3. The rotating seals required on the shaft could be a potential maintenance problem. Oscillatory mass flowmeters, while not requiring rotating seals, are slightly density-dependent in their accuracy.

11-4. Linear.

11-5. (a) Mass-flow rate, $\dot{m} = \rho V A$

A is known since the line diameter is specified.

V can be calculated with the vortex meter factor, K, and ρ can then be calculated using the equation for the pressure loss through an orifice.

$A = \pi/4 \, (3/12)^2 = 0.0049 \, \text{ft}^2$
$V = \text{freq}/K = 20 \, \text{hz}/3 \, \text{hz/fps} = 6.67 \, \text{fps}$

Rearranging Eq. (4-6):
$\rho = 2(5 \, \text{psid}) \, (144) \, (.6)^4 (.8)^2 / (6.67 \, \text{fps})^2$
$\rho = 2.69 \, \text{slug/ft}^3$

$\dot{m} = \rho V A = (2.69)(6.67)(0.049) = 0.879 \, \text{slug/sec}$

$(28.3 \, \text{lb}_m/\text{sec})$

(b) An inferential mass-flow measurement.

UNIT 12

12-1. Clamp-on Doppler flowmeter. Since the flow is highly abrasive, an obstruction-type primary element is unacceptable due to rapid wear. This eliminates many devices, leaving essentially three possibilities: a magnetic flowmeter, a correlation technique, and a sonic technique.

Any of these methods will do the job, but since high accuracy is not required, a sonic-type flowmeter is more desirable after cost comparison. Ultrasonic TOF is not practical since the signal quality would suffer from the presence of gravel. However, a clamp-on Doppler system would benefit from the heavily seeded fluid.

12-2. Positive-displacement flowmeter. The fluid properties listed rule out the use of fluid dynamic, magnetic, and turbine flowmeters. A positive-displacement flowmeter is well suited for this type of intermittent (flow—no flow) operation since its meter factor is valid from zero the maximum rated flow. The flow measurement is accurately integrated during the starting and stopping periods.

12-3. By-pass (flow-ratio technique or point-velocity measurement). In very large line sizes, when high accuracy is not required either a by-pass flowmeter or a point-velocity measuring device is an economical choice. If the flow profile is suspected to vary over the operating range such that no representative control point exists in the cross section then a by-pass system is superior.

For a sediment-laden fluid, a venturi in the main line and a target flowmeter in the by-pass is a good system since neither device is likely to collect sediment.

12-4. Target flowmeter. The fluid properties prohibit the use of fluid-dynamic or magnetic flowmeters. A head-class flowmeter is more economical than the remaining alternatives. Of the head-class flowmeters, the target meter is the best suited since it does not require pressure taps, which would be susceptible to clogging with this fluid.

Appendix C: Glossary of Flow Measurement Terminology

APPENDIX C

Glossary of Flow Measurement Terminology

NOTE: The definitions presented in this appendix are taken, when applicable, from "Glossary of Terms Used in the Measurement of Fluid Flow" (ANSI/ASME MFC-1M-1979).

absolute pressure—The combined local pressure induced by some source and the atmospheric pressure at the location of the measurement.

accuracy—The measure of freedom from error; the degree of conformity of the indicated value to the true value of the measured quantity.

anemometer—Any means used for measuring the velocity of air flow.

beta ratio—The ratio of the diameter of the constriction to the pipe diameter, β = Dconst/Dpipe.

bore Reynolds number—Calculated Reynolds number R_d using V_{bore}, ρ_{bore}, μ_{bore}, d_{bore}; also $R'_d = R_D/\beta$.

bottom contraction—The vertical distance from the crest to the floor of the weir box or channel bed.

calibration—Determination of the experimental relationship between the quantity being measured and the output of the device which measures it; where the quantity measured is obtained through a recognized standard of measurement.

centrifugal force—A force acting in a direction along and outward on the radius of turn for a mass in motion.

choked flow—The condition of maximum velocity at the minimum area section of a device corresponding to the speed of sound of the gas flowing through the device.

Coanda effect—A phenomenon of fluid attachment to one wall in the presence of two walls.

complete contraction—A combination of both end and bottom contractions in a weir.

confidence level—The probability that the interval quoted will include the true value of the quantity being measured.

contraction—The narrowing of the stream of liquid passing through a notch of a weir.

Coriolis force—Results from Coriolis acceleration acting on a mass moving with a velocity radially outward in a rotating plane.

crest—The bottom edge of a weir notch, sometimes referred to as the sill.

crest width—The distance along the crest between the sides.

critical flow—Same as *choked flow*.

differential pressure—The static pressure difference generated by the primary device when there is no difference in elevation between the upstream and downstream pressure taps.

drawdown—The curvature of the liquid surface upstream of the weir plate.

dynamic pressure—The increase in pressure above the static pressure that results from complete transformation of the kinetic energy of the fluid into potential energy.

expansion factor—Correction for the change in density between two pressure-measurement stations in a constricted flow.

flow rate—Actual velocity of the fluid medium.

flow-rate range—Range of flow rates bounded by the minimum and maximum flow rates.

flowmeter—A device for measuring the quantity or rate of flow of a moving fluid in a pipe or open channel.

flume—An adaptation of the venturi concept of flow constriction applied to open-channel flow measurement.

free flow—A condition in which the liquid surface downstream of the weir plate is far enough below the crest so that air has free access beneath the nappe.

gage pressure—The difference between the local absolute pressure of the fluid and the atmospheric pressure at the place of the measurement.

gravitational constant—A dimensionless conversion factor in English units which arises from Newton's second law (F = ma) when mass is expressed in pounds-mass (lb_m).

$g_c = 1$ in SI and cgs, units and English units when mass is expressed in "slugs."

$g_c = 32.17 \ \frac{lb_m}{slug}$ in English units when mass is expressed in lb_m.

head loss—Pressure loss in terms of a length parameter such as inches of water or millimeters of mercury.

head pressure—Expression of a pressure in terms of the height of fluid, $P = y\rho g$, where ρ is fluid density and y is the fluid column height.

isentropic exponent—A ratio defined by the specific heat at constant pressure divided by the specific heat at constant volume.

kinetic energy—Energy related to the fluid dynamic pressure, $1/2 \ \rho \ V^2$.

laminar flow—Flow under conditions in which forces due to viscosity are more significant than forces due to inertia.

linearity—The maximum percent deviation of calibration data from a zero-based straight line.

Mach number—The ratio of the fluid velocity to the velocity of sound in the fluid, at the same temperature and pressure.

mass-flow rate—The product of fluid density, ρ, full closed conduit area, A, and fluid velocity, V.

meter run—A flowmeter installed and calibrated in a section of pipe having adequate upstream and downstream length to satisfy standards of flowmeter installation.

nappe—A sheet of liquid passing through the notch and falling over the weir crest.

Newtonian flow—Fluid characteristics adhering to the linear relation between shear stress, viscosity and velocity distribution, $\tau = -\mu \frac{dV}{dy}$.

notch width—The horizontal distance between opposite sides of the weir notch.

percent of actual—Same accuracy value applies over the entire flow rate range.

percent of span—Accuracy value applies only at the maximum rated flow.

pipe Reynolds number—Calculated Reynolds number R_D, using V_{pipe}, ρ_{pipe}, μ_{pipe}, and D_{pipe}.

potential energy—Energy related to the position or height above a place to which fluid could possibly flow.

precision—Or repeatability, a random error caused by numerous small independent influences which prevent a measurement system from delivering the same reading when supplied with the same input value of the quantity being measured.

primary device—The part of a flowmeter which generates a signal responding to the flow from which the flow rate may be inferred.

proving—Determination of flowmeter performance by establishing the relationship between the volume actually passed through the meter and the volume indicated by the meter.

pulsating flow—A flow rate that varies with time, but for which the mean flow rate is constant when obtained over a sufficiently long period of time.

quantity meter—A flowmeter in which the flow is separated into known isolated quantities which are separately counted to determine the total volume passed through the meter.

random error—Precision or repeatability, data that deviate from a mean value in accordance with the laws of chance.

rangeability—The ratio of the maximum flow rate to the minimum flow rate of a meter.

repeatability—See *precision or random error*.

reproducibility—A measure of the characteristic nature of any flowmeter to be taken off the shelf, installed in a flow line and deliver a specified performance level.

resolution—The error associated with the ability to resolve a flowmeter output signal to the smallest measurable unit. For example, only ±1 pulse is measurable in any pulse output device.

Reynolds number—The ratio of inertia and viscous forces in a fluid defined by the formula $R_e = \dfrac{\rho V l}{\mu}$.

secondary device—Part of a flowmeter which receives a signal from the primary device and displays, records, and/or transmits it as a measure of the flow rate.

slip—A term commonly used to express leakage in positive-displacement flowmeters.

spurious error—Errors due to instrument malfunction or to human goof-ups.

stagnation pressure—The pressure of a fluid one would obtain if one could bring a flowing fluid to a standstill (to rest) isentropically (without any energy loss).

static pressure—The pressure of a fluid that is independent of the kinetic energy of the fluid.

static weighing—A method in which the net mass of liquid collected is deduced from tare (empty tank) and gross (full tank) weighings respectively made before the flow is diverted into the weighing tank and after it is diverted to the by-pass.

steady flow—A flow in which the flow rate in a measuring section does not vary significantly with time.

stilling basin—An area ahead of the weir plate large enough to pond the liquid so that it approaches the weir plate at low velocity, also called weir pond.

Strouhal number—A nondimensional parameter defined as: $S = \dfrac{fh}{V}$, where f is frequency, V is velocity and h is reference length.

submergence—The distance measured from the crest level to the downstream water surface when the flow is submerged, i.e., no air is contained beneath the nappe.

suppressed weir—A rectangular weir in which the width of the approach channel is equal to the crest width, i.e., there are no end contractions.

systematic error—That which cannot be reduced by increasing the number of measurements if the equipment and conditions remain unchanged.

total pressure—See *stagnation pressure*.

transitional flow—Flow between laminar and turbulent flow; generally between a pipe Reynolds number 2000 and 7000.

true mass flow—A measurement that is a direct measurement of mass and independent of the properties and the state of the fluid.

turbulent flow—Flow in which forces due to inertia are more significant than forces due to viscosity and adjacent fluid particles are more or less random in motion.

uncertainty—The interval within which the true value of a measured quantity is expected to lie with a stated probability.

unsteady flow—A flow in which the flow rate fluctuates randomly with time and for which the mean value is not constant.

volume flow rate—Calculated using the area of the full closed conduit and the average velocity in the form, $Q = V \times A$, to arrive at the total quantity of flow.

weir—An open-channel flow measurement device analogous to the orifice plate-flow constriction.

weir pond—See *stilling basin*.

working pressure—Or flowing pressure is the static pressure of the fluid immediately upstream of a primary device.

working temperature—The temperature of the fluid immediately upstream of a primary device.

Index

INDEX

Absolute pressure	30
Accuracy	34
Additive energy	9, 18, 22
Additive energy approach	157
Allen sale velocity method	185
Alternating current type—AC	18, 159
Anemometer	138
Angular momentum mass flowmeter	214
Archimedes principle	129
Area flowmeters	97
Area meter	13
ASME discharge coefficient	67-69
Axial transmission	165
Bellows meter	131
Bernoulli equation	47
Beta ratio	47, 64
Bottom contraction	111
By-pass	183
By-pass flow rate ratio	198
By-pass flowmeter	196
By-pass technique	195
Calibration	38
Calorimetric	170
Calorimetric flowmeter	20, 21, 174
Capillary tube	100-102
Centrifugal force	87
Cippoletti	109, 116
Clamp-on	163, 167
Classical venturi	73-75
Coanda effect	17, 143, 149
Complete contraction	111
Compressible flow	45, 50
Concentric orifice	62
Conditions—operating	225, 226, 231
physical	225, 226, 232
Conductivity	162
Confidence level	35
Conservation of energy	45
Constant current type—DC	159
Contraction	111
Control	223
Conventional type	13
Coriolis	208
acceleration	208
flowmeter	27
force	208
Corona discharge ion deflection flowmeter	194
Cost—relative	230
purchased	229, 236
installation	229, 236
operating	229, 236

maintenance 229
Cost considerations 229
Crest 108-113
Crest width 108-113
Critical flow 206
Cross-beam 166
Cross-correlation flowmeter 190-191
Cup anemometer 138
Current-type flowmeters 14, 15, 129, 136
Cylinder and piston 100, 102
Cylindrical hot-film sensor 172

Darcy formula 55
Deflection 183, 192
Density of water 65
Differential pressure 31
Differential-pressure producing 11
Direct current type—DC 18
Discharge coefficient 50, 64
Doppler 19, 20, 163, 167
Drawdown 111
Dry gas meter 130
Dye marker 185, 186
Dynamic pressure 30, 31
Dynamic weighing 38

Eccentric orifice 63
Elbow meter 13
Electro-motive-force 158
Energy 10
Energy additive 227
Energy extractive 227
End contractions 111
Existing tags 188
Expansion factor 51, 52
Externally heated 177
Extractive energy 9, 11, 17

Faraday 18
Faraday law 157, 158
Flow coefficient 50
Flow rate 29
Flow rate range 32
Flow straighteners 29
Flow direction measuring pitot probe 96
Flow marker 183
Flow ratio technique 195
Flowmeter selection 223, 228
Fluid-dynamic-type flowmeters 14, 16, 129, 143
Fluidic flowmeter 149
Fluidic oscillator 16, 17
Fluidic deflection flowmeter 192, 193
Flume 14

Free flow 111, 118

Gage pressure 30
Gas law 206
Gas marker 185, 186
Gas measurement 206
Gas turbine meter 141
Gear meter 136
Gyroscopic mass flowmeter 211
Gyroscopic precession 210

Hagen-Poiseuille law 55
Head 111-115
Head class 12
Hear pressure 31
Head-class, pulse-class device 205
Heat marker 185
Heated grid 175
Herschel venturi 73-75
Herschel, Clemens 61
Hot-film cone probe 172
Hot-film flush-mounted probe 172
Hot-film probes 172
Hot-film wedge probe 172
Hot-wire probes 171

ISA 1932 nozzle 79, 80
Impact tube 95
Incompressible flow 45
Induction 183, 186, 187
Inferential mass flow 225
Injection 183, 185
Ion deflection flowmeter 194
Ionic marker 187
Isentropic exponent 51

Kiel probe 94
Kinetic energy 10, 46

Laminar 33, 225
Laminar flow 52
Laminar flow equation 55
Laser Doppler Anemometer (LDA) 184, 188, 189
Laser Doppler Velocimeter (LDV) 184
Lenz 157
Linear-resistance meter 13
Linear-resistance flowmeters 100
Linearity 37
Liquid-sealed drum type 130
Lobed impeller meter 136
Long-radius flow nozzle 77, 78
Lower head 118

Mach number 32
Machined converging cone venture 73-75
Magnetic class 18, 157
Magnetic flowmeter, principle 158
Magnetic flux 158
Magnetic marker 187
Mass-flow measurement 203
    inferential 203, 204
    true 203, 204, 208
Mass-flow rate 29
Measurement
    mass 224
    rate 224
    volume 224
Meter run 28
Multibeam contra-propagating 166

Nappe 111
Newton law of friction 53, 54
Notch width 110
Nozzle 11
Nozzle-venturi, critical 207
Nuclear Magnetic Resonance (NMR) 187
Nutating disk 14
Nutating disk flowmeter 134

Open-channel measurement 107
Open-channel type 14
Orifice 11
Orifice plate types 62
Orifice and plug 97, 98
Orifice, balanced-bridge 217
Oval gear meter 14, 15

Palmer-Bowlus flume 117, 122, 123
Parobolic discharge flume 117, 123, 124
Parshall flume 117, 119-122
Percent of actual 34, 35
Percent of span 34, 35
Performance 227, 228
Performance considerations 229
Pipe elbow flowmeter 87-89
Pitot probes 13, 92-97
Pitot-static probe 94, 96
Pitot-venturi 94
Polarization 157, 159
Porous plug 101
Positive-displacement type 14, 129
Potential energy 10, 46
Powered flowmeters 157
Precessing vortex flowmeter 148
Precision 36
Pressure head 47

Pressure loss 32
Pressure loss for head-class devices 79
Pressure-averaging pitot type 94
Pressure-differential-type mass flowmeter 216
Price meter 137
Primary device 27, 28
Process control 224
Profile regulator 29
Propeller meter 16, 137, 139
Proving 38
Pulsating flow 33, 34
Pulse-producing class 14, 129
Pulsed-DC type 161

Quadrant-edge orifice 63
Quantity meter 30

Radiation ion deflection flowmeter 195
Radioactive marker 185, 186
Random error 36
Rangeability 32
Rate meter 30
Reciprocating piston meter 132
Rectangular notch 109-115
Reflected beam 166
Repeatability 36
Reynolds number 32
    Pipe 66
    orifice bore 66
Rotameter 13, 98, 99
Rotary piston meter 133
Rotary-vane meter 135
Rough-cast venturi 73-75
Rough-welded sheet metal 73-75

Salt marker 185
Seal
    positive 129
    capillary 129
Secondary device 27, 28
Segmental orifice 63
Sill 110
Sing-around technique 165, 166
Single-rotor mass flowmeter 216
Sinusoidally heated 177
Slip 129, 130
Smiles per dollar 223, 229
Sonic class 18, 157
Special head-producing flowmeters 87
Special techniques 183, 227
Special type 13
Spurious error 36
Stagnation pressure 30, 31, 47

Static pressure 30, 31
Static weighing 38
Steady flow 33
Stolz equation 70
Strouhal..17, 144
Strouhal number 146
Submerged flow 111
Submergence 118
Suppressed weir 111
Swirl meter 149
Swirl remover 29
Systematic error 36

Tagging 183, 185, 187
Tapered tube and float 98, 99
Target flowmeter 90-92
Target meter 13
Theoretical velocity 49
Thermal class 18, 20, 157
Thermal flowmeters 170
Thermo-anemometer 20, 21, 170
Time-of-flight (TOF) 19, 163
Total energy 46
Total pressure..30, 31
Transitional flow 33
Trapezoidal notch 109, 110
Triangular...109, 116
True mass flow 225
Turbine flowmeter 140, 141
Turbine meter 16
Turbine meter performance 142
Turbulent...225
Turbulent flow 33, 55
Two-diaphragm slide valve meter 130, 131

Ultrasonic flowmeter 163
Uncertainty 35
Upper head 118
Unsteady flow 33, 34
Upstream pipe requirements 72

V-notch 109, 110, 116
Vane anemometer 39, 140
Vane flowmeter 90
Velocity head 47
Velocity of approach factor 49
Venturi 11
Venturi nozzle 74, 76, 77
Venturi tubes 73, 216
Venturi, Giovanni Battista 61
Viscosity of water 65
Volume flow rate 29
Volumetric method 38

Vortex flowmeter 144
Vortex formation 143
Vortex meter 16, 17
Vortex precession 143, 148
Vortex sensors 145
Vortex shedding 143
Vortex performance 148
Vortex viscosity effects 147

Weighing method 38
Weir 14
Weir box 113
Weir dimensions 113, 114
Weir pond 111
Weir size guide 112
Wetted transducer 163, 167
Woltman meter 15
Work 46
Working pressure 31
Working temperature 31